SETTING THE HOOK

A DIVER'S RETURN TO THE ANDREA DORIA

PETER M. HUNT

Copyright © 2011 Peter M. Hunt
All rights reserved.

ISBN: 1453734201
ISBN 13: 9781453734209

...Keep Ithaka always in your mind.
Arriving there is what you're destined for.
But don't hurry the journey at all.
Better if it lasts for years,
wealthy with all you've gained on the way,
not expecting Ithaka to make you rich.

Ithaka gave you the marvelous journey.
Without her you wouldn't have set out.
She has nothing left to give you now.

And if you find her poor, Ithaka won't have fooled you.
Wise as you will have become, so full of experience,
you'll have understood by then what these Ithakas mean.

From Ithaka. C.P. Cavafy: Collected Poems,
Translated by Keeley and Sherrard,
© 1975 Princeton University Press.
Reprinted with permission.

ACKNOWLEDGEMENTS

Many people helped in the writing of this book with ideas and inspiration, especially my parents, brother Chris, and sisters Sarah and Eliza. Long time friends Steve Bielenda, Hank Keatts, Chris Dillon, and Janet Beiser provided invaluable assistance in refreshing my memory and reviewing facts. Editing help came from my nephew Henry, and friends Rick, Chris, Steve, Dave, Susan, Dennis, and Leslie. A special thanks to Bradley Sheard for the use of his excellent underwater photos.

Most importantly, here's to the *Golf Clappers:* Don, Craig, and Gary who made the memories and book possible. We're having some fun now.

Dedicated to my wife, Laurie, and children, Emily and Jared, who inspire, encourage, and endure - thank you and love always.

PROLOGUE

Three Decades of Memory

Spring of 2011, Whidbey Island, Washington

The involuntary rigid curl of the right toes is the first clue that the ankle is beginning to freeze up in a familiar twist, and with the knowledge that in a matter of minutes my confident stride will become a dragging shuffle, comes also the realization that it can't be ignored any longer. The regulator slips from my hands and drops to the workbench in my fumble for the small pill container on the key ring. Slowly, carefully, pull out a "big orange" and a "blue," bring them to my mouth and swallow the pills mechanically without water. Turn gently to find the plastic grooves, there; now screw the cap back on, feel the ache of fingers searching for the break in trouser's pocket, and finally, push the keys down to the bottom so they won't fall out, but carefully, try not to pinch the sensitive fingertips. Breathe deeply, concentrate, extend the stomach to allow the lungs to fully expand as if diving; use all the oxygen efficiently, focus on relaxing. It is just like diving, the greater the stress, the more critical that the body be allowed to relax. The right ankle starts to loosen, the toes lose their curl, and the leg is partly mine to control once again.

Done with the now familiar ritual and back to work, my hands again attempt to cinch down the cable tie securing the rubber mouthpiece to the regulator, and this time, perhaps because of the break in routine, my effort is successful. I know my body will soon return to the most recent

SETTING THE HOOK

afternoon normal, with nearly unimpeded gait, but only at the cost of a gentle full length sway, with head rolling in slow rhythm, shoulders and back taking the neck's lead and following the gentle swoons with buffered exaggeration. When the pill dosages come too closely, about noon, so also comes the inevitable trade off. To move uncontrollably or not to move at all; that's the question and the pills can no longer postpone the answer. The simple job of securing the rubber mouthpiece to the regulator now complete, my eyes slowly scan the workbench and settle on the brightly colored photograph of the *Wahoo* racing head on, full speed through the surf, recklessly young in its disregard for what will come.

My thoughts comfortably return to ten, twenty, and thirty years ago. Three decades of the shipwreck still occupy my mind - pushing and pulling, much like the disease - and it provides a focal point, a nexus of the everyday and the extraordinary where it is tempting to search for eternal truths. With a plodding doggedness the disease patiently takes the body from my control, leaving only my mind and eyes to truly own until the end, or at least that's my best hope. Maybe a few more months able to dive, perhaps a year, the only certainty is that time will prevail; that each new "normal" will come. But for now, I breathe. I breathe deeply with the cherished clarity of what is.

With the regulator shoved deep into the dive bag, my now confidently swaying body shuffles from the garage in trail of my young son barely visible beneath bags of dive gear. Three decades of memories return.

***

CHAPTER ONE

The Dive

*"When you're lost in the Wild, and you're scared as a child,
And Death looks you bang in the eye..."*

Opening line from *"The Quitter"* by Robert W. Service

The First Decade: July 24th, 1983, fifty miles south of Nantucket Island, Massachusetts

The bright, morning sun made it easy to keep sight of Craig Steinmetz pulling hand-over-hand down the taut anchor line thirty feet below. Other than Craig and his double yellow scuba tanks, there was nothing else to see. Surrounding my visual universe tiny plankton particles sparkled in the light, but the white nylon anchor line was the only tangible reference point. Nothing more above, nothing below - only water. The nearest land, Nantucket Island, was fifty miles away, and the closest mainland, Montauk Point, New York, was almost one hundred miles to the west.

I scooped my left arm back and took hold of the two tank pressure gauge hoses where they snapped together and terminated in a plastic instrument console. The array of rigid dials was reassuring: the pressure activated bottom timer was working, the small second hand sweeping, the minute arrow still waiting to click the stop watch over to the first minute of the dive. The

SETTING THE HOOK

depth indicator's black arrow crept past twenty-five feet. The instrument console fell away to dangle at my side as I reclaimed a two handed grip on the anchor line, tense in the current, and resumed my slow, steady kicking.

Taking comfort in Craig's form below, my eyes strained to pierce the darkening waters for a glimpse of our destination. Impatiently, I reached again for the instrument console while jutting my jaw every few seconds to clear the building pressure in my ears; seventy-five feet, much too soon to see anything. Breaking a threshold of sudden cold signaled what had to be the final thermocline, the well defined demarcation between the warmer surface waters and the chilled depths. The current lessened with depth, making it possible to use the anchor line as merely a guide, and a two handed grip was no longer required to avoid being swept away. The next time I held the instrument console it stayed in my left hand: 110 feet, 120 feet, the black arrow began its second clockwise rotation around the luminescent dial backing.

I became profoundly conscious of my breathing, slow, deliberate near full inhalations; it took practiced concentration to automatically extend each exhale, to control each breath. If a diver inadvertently inhaled to full capacity and ascended, quite possibly only a few feet in the shallows, the expanding air in their lungs would have no ready exit. The air bubbles could escape to myriad places in the chest cavity, all bad, some fatal. The only way to ensure against an "embolism" was to never hold your breath underwater.

Stretching each molecule of oxygen was a subconscious fight against the inexorable draw of gas from the two pressurized tanks on my back, keeping me constantly attentive to the deep seated knowledge that each sparkling exhale of bubbles from a regulator was one breath gone forever. I released the anchor line completely, shifted body angle slightly to the horizontal, and kicked free-floating directly into the diminishing current to stay within feet of the rope. With eyes locked on the anchor line I felt for the two additional regulators hanging at my shoulder, pulled them into view, and a quick glance confirmed both were easily accessible and ready to use. Each of the three regulators had a distinguishing feature - the shape or material of the regulator housing - to clearly indicate which tank was in use. I carried two full-sized, eighty cubic foot tanks and one small backup "pony bottle," an old refurbished fire extinguisher, for use as an emergency backup and a limited reserve of air.

The nagging discomfort along my spine dissolved once my fingers wrapped back around the security of the anchor line, the tenuous guide between the adventure of exploration below and the safe haven of the Research Vessel *Wahoo* bobbing in the ocean above. Craig was slowing his descent and our separation had narrowed to about twenty feet, but it was difficult to tell the real distance through the shifting ripples of current-swept plankton moving against the infinite background of darkening waters. Every few seconds a bunched beard sticking out from his neoprene hood came into view as he turned to catch sight of me in his peripheral vision before redirecting to the empty below. Back to the console's depth gauge - 135, 140 feet - nothing yet, still too soon.

The low morning sun could no longer compete with the depth and it became ominously dark. I thought about unsnapping the light at my side, but opted to keep my hands where they were, on the safety of the anchor line and the reference of the console gauges. The distance to Craig narrowed further in the gloom, and he was only ten feet below when the outline of an enormous structure began to fill and then expand what I had thought was the limit to visibility. A mix of unadulterated thrill and an ancient, instinctual, panicky dread of the unknown shot through me with fleeting intensity; at 145 feet deep, the shadow of the *Andrea Doria* appeared.

In the dull, thickly greenish soup of minute sea life it was initially difficult to imagine her as a ship. The 697 foot *Andrea Doria* lay with her starboard side nearly flush on the sand 240 feet underwater. All that could be seen was the vast, slightly rounded port side hull, overgrown with anemones and sculptures of rusted steel, making it look vaguely like a man made ocean bottom. The *Wahoo's* anchor chain was shackled into an exposed beam, where storm and current over time had displaced a steel plate. The beam was totally covered in alternating patches of bulbous, protruding anemones, and fresh rust rubbed visible by the chain. Settling my

SETTING THE HOOK

knees onto the *"Doria"* next to Craig, I surveyed the murky dark for a reference. Immediately ahead were the gutted openings of the Promenade Deck, where most of the windows had collapsed into the debris of the deteriorating wreck. My eyes scanned right, to where the hull fell off toward the keel. It was still too dark under the low morning sun to see the dim outline of the hole that Peter Gimbel's expedition of two years prior had cut in the *Andrea Doria's* side.

I glanced down to check the air in my twin main tanks. The gauge in the console read 2,300 pounds per square inch, and the second tank gauge - secured to the console with a bungeed snap - read a full, 3,000 psi. Reaching for the regulator from the full, 3,000 psi tank, I released the "descent" regulator from a gentle bite, and set my teeth on the new regulator's mouthpiece. The breaths came with only a little resistance from the depth, but now had the distinctively tinny taste of air at 170 feet. I reached for my light, flicked it on, gave a nod of "ready" to Craig, and we swam side-by-side in the direction leading to "Gimbel's Hole" fifty feet away. We had been on the *Andrea Doria's* hull for twenty seconds.

I floated weightless above the behemoth structure, equally in awe and filled with a sense of absolute empowerment. Completely free to swim in any direction, each effortless twist of course over the encrusted hull was my unimpeded choice. Every option was new to me; there was so much to take in, to see, to experience. I was twenty-one years old, the possibilities were endless, and an entire life lay before me to explore.

This was our sixth dive on the wreck and the second expedition to the *Andrea Doria* this year. Incredible as it would have seemed a month ago, we were in a semblance of a routine. Our goal was to penetrate the wreck's interior through Gimbel's Hole, then drop down another thirty-five feet and swim to the pantry where a veritable treasure trove of fine china was buried under silt deposited by years of underwater current. The dishes, which had been staged in the pantry for service in the First Class Dining room, had two distinct designs: one elaborate and ornamental, and one elegant in its simplicity. The decorative Asian design, hand painted by Richard Ginore, depicted men and women set in nature amid myriad detailed flourishes and bright colors. The "simple" dinnerware was rimmed with a braid of red with gold inlay. All sported the trademark crown and imprint of the *Andrea Doria's* Line, "Italia."

Craig reached the darkened abyss where Peter Gimbel's 1981 commercial salvage operation had torched free two sets of four foot wide steel Foyer doors. We kicked out just past the small rectangular opening, hesitated for a moment hovering weightless over the foreboding blackness, and then dropped fins first away from the light. Leaving the comparably bright outside hull for the inky blackness of Gimbel's Dish Hole inevitably instilled the unsettled feeling of being twice removed from the sun. We left the outside world above at 170 feet, lowered our eyes, and scanned depth gauges and the guts of the ship with alternating sweeps of dive lights that barely penetrated ten feet into the black.

The sound of my breathing changed and a subtle hum filled the edges of my consciousness. The bubbles in each exhale made nitrogen narcosis induced noises, distinct, crisp little pops, and joined an increasingly busy background clamor of dull buzzing and unidentifiable random clanks. Some sounds were real noises, but were they all? Which ones were merely tricks of the mind? It was impossible to tell. More than distracting, the sounds were mesmerizing, softly seductive and getting louder, hypnotizing in tone until practiced concentration spurred on by a deep-seated fear of losing mental control sharpened my focus. Why was I here? I forced the question again and again, disciplining my mind to concentrate and keep me moving toward my goal, slowly, deliberately, but with excruciating purpose; I must not forget my purpose. Do not lose focus to the false comfort of surrounding noise, do not relax, and above all, do not sleep. This mental battle was normal with the expectation of being a little "narced," but that did not mean that it took less than full concentration of effort to overcome the sedative effect. Breathing air with its pressurized nitrogen at this depth was tantamount to diving drunk. The disorienting darkness, lack of an intuitive visual reference, exertion, and cold were all part of the package for any penetration dive into the *Andrea Doria*, and each factor changed a diver's actual physiology and steadily increasing the threat of nitrogen narcosis.

SETTING THE HOOK

At 205 feet, we each began slow flutter kicks to stop a further descent into the void, pumped air into buoyancy compensators and dry suits, and hung in neutral buoyancy. Two hundred and five feet defined the location of the proper corridor. We had dropped down though Gimbel's Hole facing aft and maintained that orientation in the descent to avoid venturing inadvertently into the wrong passageway. At 205 feet and looking aft, we would be facing the correct opening, the one leading to the china.

Debris littered the mouth of the corridor, and with rotted-out bulkheads above and below, the passages looked similar in a dive light's narrow beam of artificial brilliance. Behind us, toward the bow, was the corridor to the ship's chapel. On my first dive on the *Andrea Doria*, only two weeks earlier, I had left my buddies in the "Dish Hole" for a few minutes to explore. My thirty foot swim toward the chapel - which had partially collapsed deeper into the ship - yielded a fantastic prize; a four-inch, oval jewelry box made of silver and covered in leather. The jewelry box sported a smooth, blue lapis stone on the flowing lines of its slightly raised cover. The hinge had deteriorated years ago and all four of the small legs had fallen off. I was lucky enough to find two of the legs lying next to the jewelry box in the shallow layer of silt. The prevailing currents had kept this passageway relatively clear of sediment. Not so in the Dish Hole, where the depth of silt was measured in feet.

After putting the jewelry box in my mesh goodie-bag and rejoining my buddies, we had swum to the anchor line ready for the ascent. During that lengthy decompression I was overcome with the contrasts of the dive: the grandeur of the *Andrea Doria*, her sheer size, the organized clutter of the vast, deteriorating hull, and the finely detailed china, many without a chip and still shiny under the rub of a glove. But as I reflected back on that first *Doria* dive, hanging in the current while decompressing on the anchor line, watching my air supply go down and my depth go up excruciatingly slowly, it was the mysteries of the jewelry box that held my thoughts. How had it gotten to that corridor? It was clearly a valuable item to whoever owned it in 1956 when the *Andrea Doria* sank. Under what circumstances of panic or controlled urgency had the jewelry box been lost?

THE DIVE

Enthusiasm for the *Doria* did not diminish after the first dive; instead the thrill grew exponentially. There were always more questions, more to explore, more to experience. All that could temper the rush was a healthy respect for the wreck, at least in any prudent diver. There was plenty about diving the *Andrea Doria* to concern even the most experienced adventurer, and those who refused to respect her did so at tremendous risk.

My eyes strained to pierce the shadows of the Dish Hole and concentrate, but my thoughts continued a narcosis-induced drift. Craig and I had swam past a typewriter lying isolated in the mud at the corridor entrance several weeks earlier, left untouched by the handful of divers who had previously dared to visit the spot. Craig was a bearded thirty year old ex-biker who garaged his Harley after one too many highway spills. Craig also happened to own a business machine store in Huntington, New York, where the principal commodities were typewriters; we could not let this one get away.

Two weeks earlier Craig had corralled me into helping him hump the heavy typewriter up out of the shaft and onto the side of the *Doria*, where in the better lighting conditions and clear of protruding beams and debris we could more safely attach a lift bag and send it to the surface. At 205 feet deep, with the drag of twin tanks, dry suits, a thirty-five pound weight belt, and all kinds of lights, hammers, and crow bars hanging from our waists, it was no easy feat hauling the worthless hunk of keys and rollers straight up thirty-five feet. We lived for pranks though, and the thought of what a good one it would be to display the typewriter in Craig's store compelled us to try. By the time we each got a hand on the lip of the square cut out at the top of Gimbel's Hole, my breathing was so labored that I felt dizzy, which is a pretty serious situation when breathing compressed air 170 feet underwater. We summoned a final effort and heaved the typewriter onto the *Doria's* side. Suddenly overpowered by the drunken narcosis, it was all I could do but to settle motionless on my knees, eyes closed, and try to control my breathing.

"Deep-water blackout" is a somewhat technical term for losing consciousness underwater. It's caused by a buildup of carbon dioxide, primarily from working and breathing too hard, and if on a stage decompression dive, with multiple stops at different depths required to vent the nitrogen in the bloodstream prior to surfacing - like an *Andrea Doria* dive - it is almost always fatal. A dive buddy can drop your weight belt and send you to the surface, but if you don't drown in the ascent, then you will still be so severely "bent" by the bubbling nitrogen in your bloodstream that death is likely anyway. It took me about thirty seconds, but my head began to clear. I helped Craig finish tying off the lift bag to the typewriter, filled the heavy-duty balloon with air from both our regulators to share the loss of breathing gas, and watched the typewriter disappear as it raced to the surface. Two weeks later the typewriter sat in a fish tank filled with fresh water in the store's front window with the label: "We make house calls… anywhere."

CRAIG STEINMETZ ON THE *WAHOO*'S SWIM PLATFORM WITH THE RECOVERED TYPEWRITER. FROM THE AUTHOR'S COLLECTION, PHOTOGRAPH TAKEN BY STEVE BIELENDA (1983).

THE DIVE

I pushed back the memories without breaking my stare into the undisturbed blackness of the Dish Hole. Refocusing to the task at hand, I turned on my spare light, pointed it directly into the blackness, and set it firmly at the edge of the passageway. It was easy to become disoriented between the eye-squinting visibility and mind-bending narcosis when trying to leave the Dish Hole for the *Doria's* outside hull, and more than one diver over the years would swim right past the shaft leading up to freedom. Hanging neutrally buoyant, it was difficult to tell that the "floor" dropped away and that it was time to look up for the dim, square of twilight thirty-five feet above; the shadowy outline was the only known exit from inside the Foyer Deck. Fortunately, those divers who failed to look up at the right moment and overshot the shaft eventually realized the mistake before exhausting their air supply swimming too far forward toward the *Doria's* bow. When we reached the spare dive light nearing the end of our dive we would have the reassurance of knowing that it was time to look up for the dim square of sun and freedom.

Craig went first and finned carefully down the passageway, staying low in the corridor to avoid the bulk of menacing cables strewn about and eager to catch a diver in a death grip. It is preferable to tie off a penetration reel to use as a guide when entering some wrecks. The potential problem with relying on the thin, nylon cord to find the way out of a shipwreck is twofold. Unlike a cave, where penetration lines were the hard and fast rule, a wreck's numerous sharp, metal edges can cut a line in two. There are also creases between steel plates where a line can migrate, but a diver could never fit. I had once tried to follow a penetration line inside a well known wreck only to run into a steel wall after the line had pressed into a crease between collapsed bulkheads. The way out was only ten feet to the left, but with the visibility-obscuring silt stirred up, determining which direction to look was easier said than done.

For these two reasons a penetration line could never inspire the absolute trust of a diver inside a wreck. It might be helpful; it might even save your life. But you had better not count on it exclusively, because there was a very real possibility that it might not remain intact or stay routed on your entry path. A diver had to learn each part of a wreck's interior intimately, needed to recognize landmarks, but still be able to navigate to the exit completely by feel if the ubiquitous silt stirred up from artifact hunting made dive lights useless.

After our first dive into the Dish Hole, our group came to the conclusion that a penetration line was more danger than help in the confined space. The stray cables broken free over the years, scattered debris, and zero visibility that we knew would result the second we began digging for the china made a penetration line too much of an entanglement danger. Any slack in the line might find its way around wreck debris, a tank valve or another part of dive gear that was not easily reached, particularly in the severely limited time we were able to spend at 205 feet deep. It was not excessively difficult to feel along the Dish Hole bulkhead and use it as a physical reference to the shaft leading to the ship's hull and open water, at least as long as one stayed low and away from the swaying cables. We swam with short, gentle kicks above the silt and tried not to disturb the visibility-obscuring sediment for as long as possible. After about thirty feet Craig turned around, felt for the wall with his right hand, and settled his knees in the fine mud. Now when he was ready to leave, he need only put his right hand on the bulkhead - which was actually the corridor's ceiling - to gain a reference, stay low, and continue straight ahead until he saw the spare dive light.

This was my fourth dive into the Dish Hole, and I wanted to see what was further down the passageway. I decided to leave Craig to his digging for a few minutes and kept swimming. I knew the pantry led to the *Andrea Doria's* First Class dining room and wanted to take a peek. I swam past a bathtub on the "wall," mundane evidence of prior human presence looking prominently out of place with the *Doria's* new orientation on her starboard side. At about forty feet into the corridor the "ceiling" appeared to recede and I could no longer see the bulkhead above. After about seventy feet the passage opened up into a huge, light-sucking cavern. The First Class Dining Room ran the width of the ship at the Foyer Deck; toward the surface there was about thirty feet of open space before reaching the impassable barrier of the *Andrea Doria's* port side steel bulkhead, below at least forty feet to where the *Doria's* right side rested in the sand. I kicked out into the nothingness and shined my light in a feeble attempt to illuminate anything recognizable.

TUB AND FIXTURES AS THEY APPEAR IN THE *ANDREA DORIA'S* UNDERWATER ORIENTATION RESTING ON HER STARBOARD SIDE FIVE YEARS AFTER THE AUTHOR'S FIRST DIVE INTO THE DISH HOLE. PHOTO COURTESY OF BRADLEY SHEARD (1988).

SETTING THE HOOK

My weak dive light, the only one I could afford, failed to reveal anything of immediate interest. After several futile scans of the light, I slowly turned just in time to see the billowing edges of a rapidly approaching wall of suspended silt envelope me in blackness. My eyes snapped shut to keep from getting disoriented by the floating mud particles that obscured vision at inches from the glass of my facemask and tried not to move. The silt from Craig's digging had traveled down the corridor directly behind me, and in a fraction of a second my situational awareness was shattered, putting me in a dangerously vulnerable position. Suspended in the First Class Dining Room in zero visibility, probably only twenty feet away from the entry corridor, I had lost all physical or visual reference to find the passage out, the shaft up to the light, and safety.

Suddenly, a penetration line didn't seem like such a bad idea. In a near panic I swam directly into the silt before starting to drift up or down in the dining room, possibly losing the pantry passageway forever. It seemed to take too long, and through the narcotic buzzing I strained to remember how far into the dining room I had actually ventured.

My hands reached out in a frantic search for the opening in the blackness. My hands ran smack into a solid steel bulkhead. Real, deep-seated fear struck me for the first time on any dive. I pressed my hands firmly against the bulkhead, fully realizing that this single reference was the only possible clue to an escape. What the hell had happened? It took all my effort to fight off the demon of panic at the edge of my self-control. All right, I reasoned, under the now oddly calming influence of nitrogen narcosis: I must have sunk down a few feet before turning and becoming immersed in the silt cloud. Let me follow the bulkhead up; the corridor will be there.

With both hands along the "wall," I slowly worked my way higher, struggling to keep my shit together. I had ten minutes max to get out, and there was no way that the silt would clear in that short a period of time. After that I would be running perilously low on air. If it took ten full minutes to discover the exit, I would likely have to skip my last two required stops on the way up, pretty much guaranteeing in the very best case decompression sickness and a Coast Guard helicopter ride to shore. The

THE DIVE

bulkhead gave way in front of me, my heart rate slowed and my breathing became steady; it was going to be all right. I swam forward into where the bulkhead receded. After about ten feet, I felt a renewed and intensified flutter of my heart when the "passage" abruptly stopped in a narrow "V."

What the hell was this passage; where was the way out? I froze and took a moment to control my breathing, to work my mental state back toward the tenuously thin line of rationality between paralyzing fear and out of control panic. Shit, what did I do? There was one possibility - maybe I had drifted too high when the silt hit me in the face, and the elusive hole to the outside ocean and safety was actually further below. Slowly, I backed out of the "V" and started down, descending in the dining room, careful not to lose the feel of the bulkhead in the fog of blackness and holding close my tenuous grasp of mental control. It went against basic survival instinct to go down while every tingling sense and rationale thought screamed that the way to safety was up.

Twenty feet into my descent I felt an opening. Cautiously, I pulled along the bulkhead and into the corridor, my mind racing in the hope of salvation, until a dim flicker of light became visible through the silt. I had found Craig and the way out. The entire episode could not have taken more than three or four minutes, but it would stay with me for the rest of my life; to this day I still feel the edge of panic for a fraction of a second when the light unexpectedly vanishes from a room. I bumped into Craig's darkened form, hoped that I had given the silt-producing son of a bitch a start, and settled onto my knees to spend a couple minutes looking for china before we had to leave.

My hand found the wall and I made certain to face the correct direction toward the exit shaft. I unhooked my mesh goodie bag, opened it up, and plunged my left hand into the stirred up silt and searched for the dinnerware by feel. A few minutes later, Craig tapped me on the shoulder and it was time to go. After a short swim we emerged from the silt cloud; the slight current had kept most of it to our backs. Being able to see ten to fifteen feet into the blackness with a dive light was always a cherished treat after the zero visibility of the Dish Hole.

SETTING THE HOOK

I retrieved the spare dive light left behind as our signal to look up for the exit from the Foyer Deck and switched regulators back to my second tank. We floated up to the dimly lit outline. Once clear of the darkness and swimming on the *Andrea Doria's* hull at 170 feet deep, I began to feel at ease. Our slow ascent up the anchor line to our decompression stops, our final obstacle before surfacing, was almost relaxing - we each had enough air for the forty-five minutes at the relatively shallow depths for decompression, or "hanging." We had spent twenty minutes total on the wreck.

I didn't get many pieces of china that dive, maybe two or three, but I felt good about coming back with the most important thing to return with from any dive - my ass.

THE DIVE

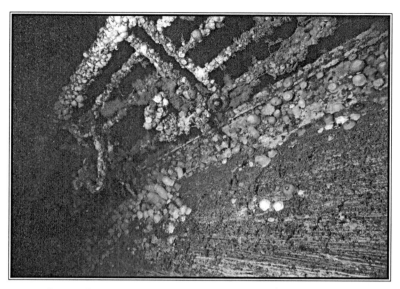

ANDREA DORIA PROMENADE DECK RAIL ENCRUSTED WITH SEA ANEMONES. PHOTO COURTESY OF BRADLEY SHEARD (1989).

FIRST CLASS CHINA RECOVERED BY THE AUTHOR AND DIVE BUDDIES. FROM THE AUTHOR'S COLLECTION, PHOTOGRAPH TAKEN BY STEVE BIELENDA (1983).

* * *

CHAPTER TWO

The Search

The Second Decade: Seventeen Years Later - June 8th, 2000, Whidbey Island, Washington State

The Park Ranger's best guess was that the boy fell in Deception Pass at about three o'clock the previous afternoon. I shifted the SeaRay smoothly into reverse and the boat left the slip, leaving me free to contemplate the ramifications the thirty hour delay might have on recovering the boy's body. The cliff face where he had presumably fallen in - an extensive search recovered a small trinket belonging to the boy at the cliff's edge - dropped down to the spiraling tempest of whirlpools that radiated from the swift current beneath the Deception Pass Bridge. Deception Pass, at the northern tip of Whidbey Island, served as a tidal-funnel compression point in the vast network of waterways and islands that made up Greater Puget Sound. The current squeezed through two narrow channels around a small, center island and could exceed eight knots during the apex of tidal exchange.

Turning the wheel to port, I shifted through neutral to forward in one motion and joined the channel leaving Cornet Bay Marina. It would not be long before we reached the cliff face where the boy had dropped to the sea. Despite the long June day, the sun was almost completely below the horizon and there still remained twenty-five minutes before the current would begin its brief slack as the tide changed direction. I moved aside, silently turned the helm over to Chris Burgess, and started suiting up.

The chances of finding the boy's body were remote to say the least. The current had been ripping for over a day, alternating direction after each of the period's three slack tides. The current never fully stopped, but merely slowed, reversed direction, and then began to accelerate with a momentum that could quickly wrench a clinging diver off the rock wall. The entire process of current reversal took fifteen to twenty-five minutes. Our only hope of locating the boy's body was if he had gotten entangled in the floating kelp or stuck in a crevasse at the base of the cliff, and each exchange of tide lessened this already slim chance of finding him.

Ron Martin, a long time dive buddy, had called me three hours earlier to ask for help. Ron lived within sight of the State Park where the boy's search party had assembled. Surprised by the activity, he had asked the Lead Ranger what happened. The search by over one hundred volunteers had been comprehensive, with one glaring exception; in the thirty hours since the boy had been reported missing, no diver had been to the site to determine if his body was in the most obvious place - underwater. I could not figure out why. There was a contingent of divers at the Naval Air Station five miles away, and I was fairly certain the Sheriff's department had a dive team. Assuming several hours of preparation time, two slack tides had been missed - the waters in Deception Pass could only be safely dived and a reasonable search conducted during one of the brief slacks in the torrent as the current changed direction.

Ron finished donning his wet suit - he had opted to forgo a dry suit tonight for the more streamlined fit of a wet suit - and stood patiently, hood on and holding his neoprene gloves while looking over the rail in the distance. A wet suit would make it marginally easier to navigate the maze of kelp and rocky crevasses lined with razor-sharp, inch long barnacles. I started pulling on my loosely fitting dry suit; the fast moving water was a bone-numbing forty-eight degrees Fahrenheit at the surface and warmth trumped streamlined fit as far as I was concerned. The sun went completely below the horizon and I looked past Ron's shoulder at the shadowy form of the bridge, looming like a benevolent guardian 180 feet above us. No other boats were on the water and nobody was visible along the rocky cliff base or adjacent beach cove. The shore search must have been called off at dark, the park was closed, and we were alone.

Diving the Pass was tricky. The currents, cold, and often miserable visibility made it a challenge, but we dove it regularly during the short spear fishing season and were reasonably comfortable in the disorienting environment of rushing water, kelp, and jagged rocks. Ron had told the Park Ranger that we would look for the boy and try to find answers for his grieving family. There was virtually no chance that he was alive and floating out to sea, but knowing for certain that he was not stuck underwater in the Pass was all that any of us could offer. We each had kids and it was easy to imagine the parent's dreaded oppression of the unknown. We would not let another slack tide go by without looking, even if that meant diving the Pass at night.

Chris would drive the boat, circling the area ready to pick us up when the current became unmanageable. Ron and I would look for the boy. Despite the darkness, I could still feel the shadow of the bridge behind and above me, could sense the steady rhythm of dim headlights crossing the span. Chris slowed the boat and turned toward the south side of Deception Pass, taking the boat in and out of gear, trying to maintain a semblance of a stable position while we waited for the current to ease.

Ron stood on the dive platform, ready to go, save for his mask which he held in his hand. I took my time; it would be another seven to eight minutes before the current eased and I did not want to tire early. At an even pace, I picked up my buoyancy compensator - the jacket that held the aluminum dive tank - and rested the eighty pound package on the transom edge. Divers wear a weight belt to help them sink in response to the positive buoyancy of a wet suit or dry suit. The buoyancy compensator's purpose was to modulate the air in its bladder to counteract the weight belt and allow for neutral buoyancy. By adjusting the amount of air in the "BC," a diver could free float at any depth. Ron reached to steady my cumbersome outfit: buoyancy compensator, tank, and the smaller emergency

SETTING THE HOOK

"pony bottle" cylinder of air. Two regulators came off the right above a third hose for my dry suit inflator. Two additional hoses were routed to the left, one connected to the buoyancy compensator and one to the tank's pressure gauge. It looked awkward, uncomfortable, and I suppose it was the first several dives of the season until the feel of the gear came back. Dive equipment had a tendency to become a part of you after practice, not strapped on, but integral to your being.

My low-level anxiety eased with the familiar fit of the gear. I was not old: thirty-eight, but much older in the adventure of spirit compared to where I had been fifteen, even five, years ago. There was a time when I would not have given a dive like this, at night, into frigid, swirling currents a second thought. Now I admitted entirely too quickly feeling just the tiniest bit on edge.

The buffeting of the water against the side of the boat eased with the approaching slack tide. I clambered onto the swim platform to join Ron standing braced against the boat's roll, now ready to dive with mask strapped to his face. I pulled on my fins and tried to keep steady with the awkward tank on my back and the heavy weight belt around my waist. Next my mask, with the strap beneath my neoprene hood, as I had learned to wear it in what seemed to be another life almost twenty years earlier. Now, if my dive buddy's fins inadvertently slapped at my face, the mask would flood, but the strap would stay secure beneath the hood to avoid losing the mask altogether. It was easy to clear the water from a flooded mask; it was nearly impossible to find one groping blindly on the ocean bottom.

We agreed with a silent nod that this was the best the current would get. The dark shape of the cliff face was barely discernable in the twilight as Chris maneuvered the boat, spun the SeaRay uncomfortably close, and put the engine in neutral. Without hesitation Ron jumped into the water. I followed like a shadow at his finned heels while dumping air from my buoyancy compensator and dry suit, knowing that the current would be more difficult to manage on the surface. Our dive lights flicked on simultaneously as we dropped to the bottom and kicked to the shallow kelp bed twenty feet away.

THE SEARCH

It was perfect slack and our effortless movements came as a surprise. We swam through the relatively clear waters of high tide with each of our lights illuminating a path fifteen feet in front of us, both now completely relaxed; this was going to be easy. We finned around and between kelp and boulders, consciously hoping to see a form recognizable as human, secretly dreading the apparition. Rockfish hovered motionless among the blooming white anemones while small crabs scurried beneath and around each nook and cranny. The rocks were alive with sea creatures, a thriving cold-water reef brimming with relief, color, and motion. Each sway of movement in my peripheral vision transformed itself into the billow of a small boy's shirt until I focused my direct gaze and saw its true source.

I could not stop visualizing my children during our search of the crevasses, first in a masochistic attempt at understanding the pain of the lost boy's family, but then with the realization of how my children were affecting me, the caution, the hesitancy, even fear that their presence in my life inspired when I did anything even remotely dangerous. It used to be that dives far more challenging than this would barely merit a second thought. In high school and college, scuba diving had been my "thing," I lived it, it was my identity. Consumed with diving to fill my need for adventure, I took my passion nearly as far as it could go. In the early 1980s, I crewed on four successful expeditions to the *Andrea Doria*, the pinnacle of shipwreck diving. They were experiences I would never forget.

We continued to kick and search, covering the entire cliff face until it turned a corner and we were directly beneath the Deception Pass Bridge. We covered the most likely spots twice, some three times, from the surface to eighty feet deep, looking for the boy who would never be found, trying to bring closure to a tragedy that would never be fully understood. The current quickly accelerated; we had eked out almost twenty-five minutes of calm, a fortunate rarity. Realizing that further search was futile, we quickly climbed back into the boat. Chris barely had time to jam the throttles forward before the now surging waters slammed the SeaRay's hull into the cliff. The ten minute ride back to port was quiet, each of us considering the boy's family, thinking of our own children, trying to divine some reason for the tragedy. At least the parents now knew that their child, separated from them by their worst nightmare, was not settled in the murk just beyond their grasp. He was gone forever.

It was not until home, while washing dive gear in the dimly lit driveway, that I began to realize it was not just the boy's death; something else was bothering me. There had been a spark, that fleeting thrill of the intensity of a challenging dive of exploration, of unique discovery, regardless of its grim manifestations this particular night. The feeling had grown alien in the past several years. What happened? I left New York to join the Navy in 1985. My appetite for adventure was more than satiated by ten years of flying A-6 Intruder attack jets off various aircraft carriers in peace and war. But it was the year 2000; I had left the Navy more than five years earlier. My diving in the interim had been consistent, but low key: nothing terribly challenging, nothing truly different to see. My two young children - Emily at six and Jared two - made me an eager prisoner to their attention. Emily's dark hair and round, China-doll beautiful face revealed the sensitive curiosity that once inspired my own soul. The all-boy, rough and tumble brawling stance of little Jared was unabashed in its challenge of the future, of the infinite possibilities. Together, the two were a microcosm of me, as if the best of what I had experienced was wrenched from my psyche and split into a pair of wide-eyed images of my life. My dives now were rarely more than a casual afternoon jaunt, returning me home to my family for dinner. I was gone half the month as it was, flying around the country as a pilot for United Airlines.

Was this just late stage "growing up"? Because of my children, was I a different person in ways not fully understood? Had I allowed myself to grow sedentary in spirit, or was this just a mid-life crisis to "work" through? Not knowing which answer was worse, or even if they were such bad answers, the questions kept coming. The comprehension came slowly, but with growing certainty, that missing was the sense of exploration, of visiting wreck sites few to none had experienced, the youthfully arrogant rush of capturing a prize, "treasure," not necessarily valuable monetarily, but priceless in the effort and risk required to bring it to the surface. There was a well-appointed curio cabinet full of *Andrea Doria* china in my living room, and a forty-two pound brass promenade deck window resting against my fireplace downstairs. If I retrieved more "wreck trophies," it would be difficult finding a suitable place to display them in my small house. I had left New York not realizing the degree to which my opportunities to dive would be curtailed. Not that it would have affected my decision, but I never came to grips with the fact that I left the northeast, its challenging wreck diving, and the adventurous lifestyle of my youth just when things were truly getting interesting.

I decided to return to the *Andrea Doria*. It suddenly became important to see what the sixteen years since my last visit had done to her, to the people who still dove down to the wreck, and ultimately what the time had done to me. When the *Andrea Doria* was launched from the Ansaldo shipyards in Sestri, Genoa she was the pride of the Italian Fleet - sleek, luxurious, and completely outfitted with all the modern conveniences available in 1951. Like her more famous counterpart the *Titanic* over forty years earlier, the *Andrea Doria's* passengers and crew experienced the transition from pleasurable luxury to horror and despair in the blink of an eye. I began to pay attention to the magazine articles friends and family occasionally sent me about someone else's adventure on the wreck I had touched so long ago. It was interesting to see that despite the significant advances in sport diving technology of the past sixteen years, the articles still universally referred to the *Andrea Doria* as the "Mount Everest" of wreck diving. It was also somewhat troubling to note that each story highlighted the numerous diving fatalities still occurring on the wreck.

There was no disguising the hollow pit in my stomach that came from the prospect of diving the *Andrea Doria* again. A variety of factors made her what we used to call in the early 1980s a "challenging dive." I would soon learn that the year 2000 term was "extreme." The liner was far from shore, one hundred miles from the nearest mainland at Montauk Point, New York, and was in the center of two opposing, international shipping lanes. The only timely assistance in the event of a problem would have to come by

SETTING THE HOOK

helicopter. The *Andrea Doria* was in deep water: 165 feet to small portions of her most shallow structure, 240 feet to the sandy bottom.

The wreck's depth provided a range of diving challenges. There was the lengthy decompression necessary to vent the nitrogen absorbed in the body to avoid the life threatening "bends," and the danger of nitrogen narcosis (what used to be called "rapture of the deep"), which made it similar to diving while under the influence of anesthesia. Performing the most basic of tasks could become incredibly complex when a diver felt "drunk" from nitrogen narcosis. The over-expansion of air in the lungs could cause an air embolism and kill a diver if they lost their head underwater and tried to claw to the surface. The swift currents, cold waters, wind driven waves and all too common fog exacerbated the challenges.

These were the obvious dangers. Subtler were the difficulties in diving a shipwreck, particularly one shrouded in lost fishing nets seemingly eager to catch an unwary diver in a death grip. Navigating in the silt-strewn, often zero visibility, interior of a wreck required skill and calm, particularly if the diver was "narced" and disoriented. It was easy to get lost or entangled in the chaos of loose cables and lines. Every second that clicked away trying to extricate oneself brought a diver closer to drowning or an almost certain fatal rush to the surface, bypassing the necessary decompression, as the last reserves of oxygen were depleted by the body. Virtually all of an intact wreck's "treasures" would be found inside, and the temptation to go a step too far always beckoned.

Reflecting on the real dangers and the possible disasters, never mind the logistics and cost, initially inspired little more than serious doubt. But I also knew that most of my concern lay in the fact that I was not ready to dive the *Andrea Doria* immediately. The summer had yet to even begin, and the earliest I could realistically be ready to dive and chartered on an expedition to the wreck would be the following year in the summer of 2001.

The *Doria* dominated my thoughts those first weeks after making up my mind. It was not until the beginning of August that I discussed the idea with my wife, Laurie, and she reluctantly went along with the plan. If I was not quite certain why I wanted to do this, it must have confused the hell out

THE SEARCH

of Laurie. Our summers on Whidbey Island were beautiful, and we had developed a tradition of front-loading my diving in the spring and June, and then spending all free time boating with the family for the remainder of the summer. Other than cleaning the bottom of the SeaRay, I had not put on my dive gear in two months.

Growing anxious to get underwater for the reassurance of some deep practice dives, my frustration mounted as my airline schedule would not allow it for two weeks. I was on the verge of obsession with the following summer's expedition, but also began to regain focus now that I had a clear goal. I began to consume life with a newly rediscovered energy, an excitement akin to a child's vibrant anticipation of a roller coaster ride.

By the end of August I decided to make the jump and commit myself. I picked up the phone and called my friend Steve Bielenda in New York, the owner and captain of the charter boat *Wahoo* that my friends and I seemed to live on in the early 1980s. The conversation immediately slammed me back to the reality I had been subconsciously preparing for, the unbridled excitement of discovery that can only come from putting it on the line, by taking a risk.

Steve's thick, Brooklyn accent was dull in its inflection and he seemed to lack his usual enthusiasm and energy. It was a voice different from what I had grown accustomed to the past several years; it was the voice of serious reflection. The week prior, one of Steve's good friends, Tony Maffatone, had died on the *Wahoo* in a diving accident. Maffatone had been as expert as they came - a Navy SEAL in Vietnam, an inventor, a veteran diver. At the time of the mishap he was diving with his personally designed experimental equipment, but on a relatively shallow (115 feet deep) and well-known wreck, the World War One heavy cruiser "U.S.S. *San Diego.*" No one was exempt from the dangers of the alien environment when the thrill of discovery pushed a diver's limits too far.

A strange and unsettling thought crossed my mind. I was only aware of one other close friend of Steve's to die underwater - Ray Ferrara. Ray's accident had occurred 130 feet deep on the New York wreck of the steam-sailing ship *"Oregon."* This had happened several years before I actually met Steve, but it was a common link nonetheless.

Ray would fill his tanks at the dive store where I worked as a teenager. He was a hell of a nice guy, uncharacteristically nice for the east coast wreck diving crowd, and I always made sure to top off his tanks at a significantly higher pressure than their rating allowed. More was always better when it came to air. Ray was the only person during my three years of working at the dive shop who ever gave me a tip. Soon after meeting Steve Bielenda for the first time I learned of the connection, and immediately felt an inexplicable kinship to this man twenty-five years my senior.

Similar to flying, the vast majority of diving fatalities are caused by human error. The deaths of Tony Maffatone and Ray Ferrara both pointed to the most unusual reason for accidents: equipment failure, technical malfunctions. Ray's tragedy was at the dawn of the popularization of wreck diving, when it was not a mainstream diving endeavor. Tony Maffatone's death occurred at the height of a technical advance in dive gear, a revolutionary period as well for exactly that reason.

I considered how the death of two of Steve's close friends bracketed all those years, the time I was now attempting to recapture. Probably pure coincidence, but it was unsettling nonetheless.

This was real, and there was no fooling myself about the danger. I told Steve my plans and he assured me of a spot on each of the two *Doria* charters the *Wahoo* had scheduled for the following summer. I hung up the phone and thought back to the diving life our group of friends had thrived on those many years ago.

CHAPTER THREE

The Wahoo

I met Steve Bielenda in June of 1982 after hearing about his new dive boat the *Wahoo*. During college summers, I worked at Ed Betts' Island Dive Shop in Huntington and crewed on his charter boat, the "*Sailfish*," during the weekends. Early in the 1982 dive season the converted tug's tired diesel engine gave a final death rattle, and suddenly we had no way of reaching the Long Island wreck sites. Fortunately, my high school friend and dive buddy, Don Schnell, had heard about Steve and the *Wahoo* from a *Sailfish* passenger and we did not have to skip even a weekend of diving. We signed up by phone for a day trip to the wreck of the "Gloria and Delores" and drove to the Captree boat basin on Long Island's south shore to take a gander at our new vehicle to adventure.

The fifty-five foot fiberglass *Wahoo* had just moved into the commercial marina and was temporarily tied up past the refueling pumps and well away from the bulk of the weathered charter fishing fleet. It was clear from the first parking lot glance that the *Wahoo* - freshly christened and custom-designed from the keel up for diving - was going to add a new dynamic to the New York charter boat competition. With shiny twin diesels, state of the art electronics, and an experienced wreck diver for a Captain, the *Wahoo's* capabilities and amenities easily surpassed those of most competitors. The *Wahoo* could take divers relatively quickly and reliably anywhere within one hundred nautical miles of land. The *Wahoo's* spacious aft deck had plenty of room for dive gear above, and still more below under the raised "doghouse," where crew and a few passengers bunked down during overnight trips. A main deck common area and mess hall covered additional berthing in the bow. There was ample space for twenty passengers on a day trip and fifteen overnight.

The *Wahoo* at the Captree Boat Basin. From the author's collection (1982).

I remember the first time I saw Steve. Streams of divers carted and hand carried their bulky gear in the six am morning cool, while Steve strutted back and forth on the dock, occasionally answering a yelled greeting, but obviously preoccupied with some greater mission. Steve was a powerful, fireplug of a man; forty-five years old at the time, he had an inner intensity that could be felt from the other end of the dock. His extremely short, thinning hair was the rugged gray of a guy who had spent his life in the sea breeze. He had a round face with a disarming smile that one had to look past to the eyes to determine if it was a smile of pleasure, or a challenge of, "What the fuck's the matter with you?" His penetrating eyes, bull neck, and barrel chest immediately told most people that it might be a bad idea to question the motivation of the smile too deeply.

THE WAHOO

Steve Bielenda was a Brooklyn tough guy. He had clawed his way into the gas station business in the not so reputable neighborhood of Bedford Stuyvesant in the 1950s and 60s, and eventually owned three stations. He caught the diving bug in 1959 and was a scuba instructor and instructor trainer by 1962. He never graduated from high school, but had more than enough street smarts and natural intellect to compensate. He was a classic entrepreneur with a Brooklyn bent - he could be ruthless in the pursuit of a better life for his family, but he also had a much softer side that he rarely showed, and never publically.

There were only a handful of charter dive boats working the wrecks to the south of Long Island in 1982. Once the *Sailfish* broke down there were really only two full time charters that originated in New York, although several boats from New Jersey would occasionally make their way out to "our" wrecks. Steve Bielenda and Sal Arena, the captain of the "other boat," the "*Sea Hunter,*" had no love lost between them. Over the years I took the tireless competition, usually expressed as downright hatred, between the two for granted; there was absolutely nothing to be gained from getting involved in the on-going feud. Very few of us weekend crew on the fringes got in the middle of the grudge matches that occasionally played out miles at sea over a distant shipwreck. We would grumble agreement about the other guy's screwed up operation, but to this day I do not know where all the animosity really came from, other than pure, dog-eat-dog New York business competition, of course.

The friction between the powerful egos and personalities of the east-coast wreck diving crowd was certainly consistent. Where the various scuba diving instructional agencies and equipment manufacturers might subtly bad-mouth each other on a retail store-to-store level, the tension between the charter operators was far more visceral. I used to half joke that the only thing the dive stores agreed on was that their competing retailers were screwed-up. The only thing the full time charter boat operators agreed on was that they hated each other. What the hell, it was engaging entertainment at its finest. If I did not know some of the players so well, I would have sworn it was all staged. Steve Bielenda was always good to me though, and over the years I learned a lot from him, and not just about diving either. This high school dropout had common sense insights into psychology and human nature that were far more practical than I was ever going to learn in a college class.

Steve was gruff, rough around the edges (and somewhat in the middle), and fiercely protective of what he held dear: his family, his boat, his friends, and his crew. I think, perhaps, the reason he had such dramatic ups and downs over the years with so many people close to him was his naturally confrontational competitiveness in the manner of the tough school of the old neighborhood. Steve was a stubborn, dumb Polack: just ask him, or don't - he would tell you anyway.

It did not take long to realize there was no middle ground with Steve - you either loved or hated the guy, you were either with him or against him. The sheer power of his character induced more than a few of his friends to exchange their love-hate dichotomy with him over the years, some several times. Steve Bielenda was a piece of work.

STEVE BIELENDA ON THE *WAHOO*. FROM THE AUTHOR'S COLLECTION (1982).

THE WAHOO

I had only just returned from Brown University for the summer when my dive buddy Don Schnell and I scrambled last minute to get a spot on the *Wahoo* charter. Don started diving in 1976; I made my first sojourn on scuba in 1979 at the age of seventeen. Don was a quiet, hard working welder with a prominent nose that he brandished proudly. We respected his unabashed calm in the face of our ceaseless ribbing, and dubbed him with the affectionate nickname, "Hook Nosed Bastard." We called him "Hook" for short.

The wreck of the day was not particularly historic; in fact, at the time no one knew for certain the type of ship or the reason for its sinking. The "Gloria and Delores," or "G and D," was a scattered mass of deteriorating metal plates and bits of superstructure broken up over the years by the winter Atlantic storms. The broken wreckage lay in low piles along the ocean floor at 105 feet and had been discovered by a pair of fishermen, as was often the case with northeast shipwrecks. They re-christened the shipwreck "Gloria and Delores" in honor of their wives, or was it their girlfriends; the wreck had many mysteries. The steady troupe of divers that regularly visited the "G and D" had taken most of the valued artifacts, mostly portholes, but the "G and D" continued to yield treasure in the renewable resource of lobster.

Don and I kept a low profile (except for Hook's nose) while we tried to figure out the routine and pecking order on the *Wahoo*. We strapped on our twin aluminum tanks twice, trying to gauge during our two hour surface interval between dives whether Steve was just a hard-nosed son of a bitch, or a professional who took his job seriously. The only conclusions we came to that day were that Steve ran a tight operation, the *Wahoo's* twin diesels were far more reliable than the *Sailfish's* clunker, and that the crew's attitude lightened markedly once all divers were safely onboard, the anchor was pulled, and the beer started flowing.

The setup was nearly perfect. Hook had steady work welding, and I was splitting my weekdays at the dive shop and working at a halfway house for the mentally handicapped. We both had every weekend to head out on the boat and get wet. The only drawback was that being a college

SETTING THE HOOK

kid my wallet was light, and my favorite hobbies after diving, chasing (not necessarily catching) girls and drinking Budweiser with the boys, made it lighter. Fortunately, Don and I had exhibited a modicum of dive skill and diplomacy in assisting the other paying passengers during our first several trips, Steve took a shining to us, and we were offered free passage as long as we helped out as crew. We were in heaven.

Our underwater duties were limited to the two critical dives in any open water, wreck diving operation - setting and freeing the anchor, or "hook" (little "h," not to be confused with Don Schnell, big "Hook"). Steve would find the wreck of the day using the Loran, a navigational instrument that transformed radio triangulations from various shore transmitting facilities into latitude and longitude coordinates. Then he would produce a mental image of how the wreck lay on the bottom by sweeping back and forth above it with the *Wahoo's* depth sounder. As much art as science, a captain's experience and patience were essential in getting the boat lined up precisely on the wreck in the chop and swell from a hundred plus feet above.

Once satisfied the *Wahoo* was properly positioned on the two-dimensional depth sounder picture and correcting for any wind and current, Steve would let out a blast on the horn from the *Wahoo's* second story superstructure. Immediately the crew at the bow would throw the anchor and coiled line into the water. A grapnel hook with its extended steel fingers to snag a piece of wreckage was usually used, but if the plan was to drag the drifting *Wahoo* back into the wreck along the sand, then the broad flukes of a Danforth anchor might be appropriate. In rare instances of absolute calm, the only thing attached to the anchor chain might be a shackle screwed shut hand tight.

Shortly after the "hook" was dropped, another member of the crew sitting in the stern, strapped to double tanks, dry suit zipped, fins on, and mask secure would receive the verbal signal to "go." A quick shuffle forward and the diver would step through the cut out in the *Wahoo's* safety rail and without hesitation jump the five feet to the water below. Simultaneously he or she would turn in mid air toward the bow so that they could immediately begin kicking forward to the anchor line. All the air would be expelled from the diver's buoyancy compensator and dry suit while still on the boat deck. This kept the diver from wasting time on the surface and allowed a descent while finning the forty feet forward to the anchor line. Speed was crucial to avoid having the anchor torn loose from the wreckage as the captain above attempted to jockey the boat in a stable position. Upon reaching the anchor line, the diver would use it as a guide in a race to the bottom, while being careful not to yank in on the slack and possibly pull the grapnel out of the wreck.

My most vivid memories of wreck diving were of "setting the hook;" those few rushed moments descending with no visual picture below but the slack anchor line enveloped in the swirling plankton near the surface. It was always a surprise - would the anchor be sitting in the sand, possibly with awful visibility and nothing but a guess at a compass course to begin dragging it to where you thought the wreck might be? This was rarely case; Steve was better than that. The diver sent down to set the hook was almost always treated to that most spectacular vision of a man-made silhouette appearing out of the nothingness, what used to be a ship, now a reef home for the fish and lobster. The eerie, dark shadow would gradually take shape into a defined outline. Depending on the wreck it might be immediately recognizable as a ship. Usually it took more time to find a familiar feature; one had to look past the encrusted sea life covering it in a protective attempt at anonymity, an attempt to return man's intrusion back to the sea.

But this first dive of the day was not for sightseeing. The job was to secure the anchor so the paying passengers would not have uncertainty going down the line; they needed to know exactly where the anchor was set on the shipwreck. If the initial spotting of the anchor was appropriately shallow, near the top of the wreck in an easily identifiable position, all that remained was to wrap the chain snuggly around an exposed crossbeam or protruding piece of wreckage. The hook needed to stay secured until the last dive of the day, when another crewmember would descend and set it free. The anchor line doubled as the decompression line and was the only sure path back to the *Wahoo* on the surface and safety.

SETTING THE HOOK

With the hook set, the diving crewmember would blow a puff of air into a small Styrofoam cup to help accelerate it to the surface to signal the boat crew. Immediately upon spying the marker, the crew at the *Wahoo's* bow would yell, "Cup's up!" and Steve would shut the diesels down. The passengers were now free to suit up for their dive into history. Setting the hook was not necessarily easy, and it cut into the crewmember's own time to spend in exploration on the bottom, but it was a mark of pride to be chosen by Steve to set the hook - it was a singularly important task that he did not trust to just anyone.

Most of the other regular duties of the *Wahoo* crew were more mundane than dropping, setting, and freeing the hook, but in some respects just as important. The *Wahoo* was a business, Steve had jumped in with both feet, and if he did not make money his family did not eat. Steve was a shrewd guy, and it never came to this, but the pressure of running his own business in a field as tenuous and seasonal as a dive charter in New York must have been tough. Our job as crew was to give passengers the safest, best adventure for their money, and it often required diplomacy in addition to simply holding a passenger's tank while they geared up.

There was an additional key ingredient in which the crew of the *Wahoo* excelled. We knew how to have fun, and Steve was most definitely aware of this when he let so many of us help him out. Like a beer-drinking virus, our antics were contagious. If the paying passengers had a good time they would tell their friends and return. We were good for business.

Practical jokes were a free for all outlet to "get" whoever might be "getable." One had to be on their toes aboard the *Wahoo*; complacency underwater meant one might die, while on the surface it would eventually mean being the butt of someone else's practical joke. The important point was to never let on that any of it bothered you. If you did, the rest of us would circle like a school of sharks to take advantage of the momentary weakness. It was all in good fun, at least most of the time. But if the *Wahoo* crew had a motto, it was "what goes around comes around," and we all, without exception, got our turn in the barrel, and Steve was by no means immune to the pranks. He was good at it, so you had to be careful, but we definitely got a few over on him.

After the dive season was over, one cold, snowy December night, Craig, Hook, and I drove out to Steve's house to socialize. We three lived in Northport, a third of the way out Long Island from the city; Steve lived about thirty minutes further east. After arriving, we were forced to endure an endless tirade as Steve regaled us with the story of how the neighborhood kids had made a cottage industry out of stealing his street sign. The quaint, wooden "Quiet Court" banner had been simply replaced after the first incident, but a repeat theft inspired Steve and his crafty neighbors to get shrewd. The next signpost they purchased was solid steel, sunk in a foundation of concrete. They would show the little bastards, Steve explained to us again and again.

After about the fifth rendition of the story and bored to tears, we said our good-byes, climbed into Craig's black, windowless Chevy van, replete with "If this van's a rock'n, don't come knock'n" license plate holder, and started home. Craig suddenly stopped just beyond the entrance to Steve's street and threw the heavy van with its V-8 engine and bumpers reinforced for just such a contingency into reverse.

"Bang!" The rear bumper smacked the Quiet Court sign planted in cement at five miles per hour - nothing. Craig pulled forward twenty feet, slammed the van back into reverse, and punched the accelerator.

"Bang!" Steve had sunk the sign deep. "Bang!" The sign leaned over.

Two more iterations and the cement footing gave way. Hook and I jumped out and lifted the seven foot post and hunk of loose concrete into the back of Craig's van before the surrounding houses could light up to investigate the noise.

That was the easy part. Stealing Steve's street sign was simple vandalism and would not hold up to even casual scrutiny for prank evaluation. Fortunately the *Wahoo* would sit dormant pier-side until spring, and we had time to consider our options. The months went by and the Quiet Court sign lay silent in Craig's garage. Whenever we spoke with Steve we were treated to yet another rendition of how the little bastard neighborhood kids were one up on him. He was not beaten - Steve would never admit defeat - and by the time April rolled around all of Steve's friends were intimately and painfully familiar with his sign problem.

The *Wahoo* was finally scheduled to leave port for the first charter of the season. It was a beautiful weekend, and Steve let it be known that he was going to be the first to jump in the water to initiate the dive season with a bag full of lobster. Before the *Wahoo* left shore, I offered to buy Steve a cup of coffee at the concession down the dock. This was an almost unheard of offer from the broke college kid, an offer he could not refuse. While we were out of sight, Craig and Hook hurriedly humped out the eighty pounds of sign and cement hidden in a wrap of blankets onto the *Wahoo* and down to the doghouse below.

Steve was in a great mood and was indeed the first to enter the water after the hook was set. Once certain he was well down the anchor line, we got to work. When Steve finished his time on the bottom and reached his ten foot decompression stop, mesh goodie bag filled with lobsters, he was confronted eye-to-eye with his nemesis of the winter. Six miles offshore and hanging by a tightly knotted rope to the anchor line was all seven feet of sign and concrete of "Quiet Court," mocking him in silence. Steve was forced to hang onto the rope below the *Wahoo* for a full ten minutes while he decompressed, face to face with the fact that we had gotten him good.

Steve Bielenda was a big kid at heart. He wanted to have a good time, and wanted those around him to be fun. To run dive charters like this meant taking on a large number of crew and making less money. Although this verged on sacrilege to a part of Steve's spirit, the drive to enjoy life won out. There was almost always too large a gathering of Steve's friends on board the *Wahoo*, all accomplished divers, and virtually all acting as crew and going for free. The personal service given to each passenger eased the loss of revenue, but where the extra hands really paid off was when something went wrong.

CHAPTER FOUR

Tauchretter

The number of certified divers grew rapidly in the 1970s as scuba classes became more accessible to the general public. Traditionally, scuba classes had been a test of physical endurance and general abuse, all steeped in a thinly veiled and watered down version of military dive training. The military offered one of the few structured programs early on from which a civilian course of training could be modeled. Sport diving's expansion was a reflection of the retail scuba industry's recognition that the rigid, military-fashioned training of old alienated a vast portion of the populace. Men and women of all ages and abilities were taking to the water in more sophisticated, user-friendly scuba diving courses of instruction.

It was a great opportunity for all involved. The dive businesses could sell gear, the charter boats could fill, and most importantly, a large segment of the population could be introduced first hand to an experience previously only available vicariously through a Jacques Cousteau special. And it also left the decision on what to dive, and how to dive, more firmly in the hands of the individual. There was no structured "weeding out" process in training; that had been largely left behind with the military model. If a student were determined, he or she could become certified. There were no laws regarding who could dive; the industry, in an attempt to avoid the stifling regulation of the private pilot world, strenuously promoted a reasonably strict program of self-policing. Dive shops agreed to fill the tanks

SETTING THE HOOK

of only certified divers and charter boats allowed aboard only those with "official" qualifications from a popularly recognized instructional agency. Of course, no one was monitoring those agencies for quality control and standardization. Still, the self-regulation seemed to work well. The only law governing New York wreck diving was coincidental - as with high-pressure tanks used for welding or hospital oxygen, the department of transportation required the hydrostatic testing of dive cylinders every five years to prevent accidental explosions when filling. The dive industry was on its own in regard to regulation, overall an excellent deal, particularly for the independent minded adventurers who were likely to take up scuba. What pockets of divers tended to forget, or perhaps never considered, was that not all diving was the same, and that visits to the underwater world could be vastly different depending on the purpose of the dive.

Most casual divers would stick to vacations in the clear, warm waters of the tropics. Many serious divers preferred the rugged challenge of a deep, cold, dark northeast shipwreck. Some fell between, striving to attain the experience to safely explore an ocean wreck, while still at a point where the diving bug had not infected them so greatly that it was their passion. Experience, as always, is the hardest of teachers, and those who took northeast wreck diving even a shade too casually were prone to overextend their abilities. Diving, like flying, is unforgiving. Statistically there may be few accidents, but when they do occur they are often fatal. Many tough lessons were learned over the years on northeast dive boats, and usually those who regularly took part in the challenge were the best students. The greatest danger was with those who participated in wreck diving erratically, but who possessed what appeared to be impressive credentials: certification cards with bolded words emblazoned like "Instructor or Divemaster." "De-structors" we used to call them. Without an experience based respect for deep-water wrecks, the probability of a diver encountering serious trouble shot way up.

There averaged one significant diving incident a season on the *Wahoo*, one that would require a diver's evacuation by helicopter (usually Coast Guard, occasionally Air National Guard). There were many more cases where a crewmember would need to swim a line out to a panicked diver on the surface, or be required to jump in the water to help a diver close to the boat for a brief, but intense rescue. Even the serious incidents didn't necessarily end tragically; often a visit to the recompression chamber, or simply the hospital, was enough to get a struggling east coast wreck diver back in fins again. Each mishap, big or small, created pressure for the passengers, stress that could be greatly alleviated by ample crew.

Steve never had any trouble getting unpaid volunteers to crew for him; it was a badge of honor to work on a dive boat, especially the *Wahoo*. But Steve's margin was too thin to have more than one full time, paid individual on board. Her name was Janet Beiser.

Janet was introduced to Steve after taking a marine biology course taught by Professor Hank Keatts at Suffolk Community College on Long Island. Hank was a long time friend of Steve's, a world-class diver, and an avid marine historian. Janet needed a job and loved to dive. She also possessed the proper character balance to put up with a lot of shit from Steve and bounce back apparently unfazed. Janet was a big woman, perfect for effortlessly hefting tanks and weights. She wore her long, light brown hair tied roughly back, and her brusque manner and physical size often masked a natural shyness. Her refusal to take shit from anybody but Steve made her an initial curiosity to newcomers, which would quickly transform into grudging respect, and then often an admiration of her professional ability. Of course, some people she just pissed off.

Janet became Captain Janet Beiser in 1983, when she was Coast Guard licensed to operate up to one hundred-ton passenger vessels 200 miles offshore. She quickly gathered expertise between running dive trips on the *Wahoo* in the summer, and fishing charters on neighboring boats at the Captree Boat Basin in the spring and fall where she earned the nickname "Captain Cod." As of 2001 she was the only licensed female Captain to run regular dive charters to the *Andrea Doria*. She also happened to be the second women to dive the *Andrea Doria* in 1981, and the youngest to do so at twenty-one.

SETTING THE HOOK

The most popular shipwrecks visited by the *Wahoo* proved to be those with the most fascinating history and compelling story. Captains Steve and Janet ran weekly trips in the summer to several such wrecks, especially the "*San Diego*" and the "*Oregon.*" The *San Diego* was a World War One heavy cruiser that held the dubious distinction of being the only major U.S. Navy combatant to be sunk in man's first modern war. She was the victim of a mine laid by a German U-boat, probably the *U-156*, and rested completely upside down with her deck flush to the sand only six miles south of Long Island. Because the *San Diego's* deck was inverted and even with the sand, access to her interior was a challenge - going shallower in depth inside the wreck meant rising closer to the keel where there was virtually no exit.

Winter storms and the inevitable deterioration of time broke up the *San Diego* slowly, and each season new ways were discovered to access her treasures within: portholes, ammunition canisters and bullets, china, and most anything that could be identified as a piece of the historic ship. The intentional violence of the warship's sinking lent an aura of mystery and intrigue, and the *San Diego's* maximum depth of 120 feet gave great latitude to the skill level required to safely explore the wreck.

The degree of experience needed to safely dive the *San Diego* varied tremendously depending on the dive's purpose. The near novice could comfortably site see at the wreck's keel 80-90 feet deep and watch endless schools of fish swimming along the vast hull on a sunny day. Once inside the wreck, however, the level of risk ratcheted up quickly, but even this apparently clear distinction between inside or outside the wreck was deceivingly nuanced. Sticking one's head ten feet into a wide gouge in the *San Diego's* side differed fundamentally from penetrating one hundred plus feet in tenuous and incremental exploration of the deteriorating, upside down hull. In many respects the *San Diego* offered all things to all divers.

The wreck of the *Oregon* had a rich history as well. The *Oregon* was a steam-sail that sank in 1886 in a collision south of Long Island at a depth of 130 feet. In 1884 she held the honor of being the fastest vessel to ever cross the Atlantic. Nearly one hundred years later, most of her structure had collapsed and her intact boilers were the wreck's most recognizable feature. Because the *Oregon's* wreckage field lay low in the sand between 110 and 130 feet deep, there was no shallow diving option as existed on the *San Diego*. This fact alone weeded out the newest of divers. To experience a twenty-five minute exploration of the *Oregon* required a total decompression time of at least ten minutes hanging on the anchor line below the water's surface. A diver belonged to the wreck during decompression diving, and any problem encountered would need to be dealt with underwater. Surfacing immediately was not an option if one were to avoid decompression sickness, or the "bends."

Decompression is necessary after particularly deeper or longer dives due to the physiological characteristics of breathing air while the body is under pressure. Air is composed of 78% nitrogen, 21% oxygen, and small trace elements of other gases rounding out the last one percent. Pressure increases with depth; the greater the pressure of nitrogen when breathed, the more of it that enters the body through a diver's circulatory system. Once a diver is underwater, pressurized nitrogen is squeezed smaller and it is essentially pushed into the body at a fraction of its size at the surface.

After dissolving into the blood stream, the nitrogen is absorbed into various body tissues uneventfully, but becomes a potential problem when the diver begins his or her ascent to the surface. If the depressurization, or "decompression," is conducted slowly, the nitrogen has time to leave the body tissues, reenter the blood stream, and naturally find an exit via the alveoli in the lungs, where it is returned to a gaseous state and exhaled. If, however, a diver's ascent is made too quickly, and the nitrogen in the bloodstream is not afforded the opportunity to vent, it will still return to a gaseous form, only this will happen while in the diver's body. The same basic concept can be applied to a soda can. Shaking a soda can increases the internal pressure and causes a rapid decompression when opened. The result is that the carbon dioxide bubbles over the side of the can as it quickly changes from liquid to gas.

It is intuitively obvious that it would be very bad to have nitrogen bubbling in one's body. The damage could range from the mild discomfort of a skin rash to excruciating pain and paralysis or death. The more time spent at depth, the greater the amount of nitrogen in saturation and the longer the needed decompression to safely vent and avoid the "bends." The term "bends" was originally coined by Brooklyn and Eads Bridge (located in St. Louis) workers in the 19th century, men who would become pressurized in caissons while they dug out water-beds for the bridge abutments. Caissons are pressurized, watertight compartments used for below the water line work without getting wet. "Caisson's disease" was the first medical term assigned to the ailment that befell workers who emerged from the pressurized caissons with unusual symptoms. "Bends" was the apt description for this mysterious ailment that left the workers writhing on the ground in a twisted pretzel of pain. The powerfully descriptive word, "bends," fit the diving vernacular perfectly.

Decompression was necessary if one were to spend any worthwhile amount of time on the wreck of the *Oregon*. Due to a lack of intact superstructure on the *Oregon* that could serve as a visible reference, it was easy to lose track of the anchor's location, particularly in conditions of poor visibility. Without the anchor line, a free ascent at the whim of the current during decompression might be a lost diver's only option. The cadre of divers that visited the *Oregon* generally required more experience than those who explored the exterior of the *San Diego*, particularly if they were to recover her most valued prize - a porthole.

An *Oregon* porthole weighed seventy pounds. To recover a brass porthole required a hammer and a chisel to break free each of the twelve bolts securing it to the hull, a job that could take a dozen dives to complete. Only then could a "lift bag," an open ended, heavy-duty balloon, be secured to the porthole, filled with air from a diver's regulator, and raised to the surface. There were other, easier to recover, artifacts on the *Oregon* as well: pieces of china, silverware, glassware, ornamental lead emblems that once graced the ship's dining room timbers, and, as on most of the New York wrecks, there were lobsters. The *Oregon* was a treasure trove for artifact-seekers with each winter storm's churning waters exposing new prizes long buried in the sand.

The third notable wreck that the *Wahoo* regularly visited was the "*U-853*." People were generally amazed to learn that a warship like the *San Diego* was sunk by the Germans only six miles from the mainland in 1918. They were equally in awe of the fact that a World War Two German submarine rested almost completely intact only seven miles from Block Island, Rhode Island. The *U-853* Captain exercised the bad judgment to sink the coal collier "*Blackpoint*" at the close of the war after offensive U-boat operations had been officially halted (it's also possible that the Captain did not receive the cease-fire message). The *Blackpoint* was a target of questionable value; she probably would have been scrapped if not for the war. What made matters worse for the U-boat was that the *Blackpoint* was targeted only thirty miles from several war ships belonging to a U.S. Navy task force, and directly offshore from the U.S. Naval base at Newport, Rhode Island.

The *U-853* was sunk by depth charges in the final hours of World War Two. Thirty-seven years later, in 1982, the *U-853* sat upright in the sand, clearly recognizable as a submarine in 130 feet of water. This was an eerie dive. It was clear to all who visited the *U-853* that she was more than an ordinary shipwreck - she was a war grave, and the crew's remains were scattered within the wreck. Nowhere was one side of a budding controversy better illustrated than inside the cramped confines of the German submarine.

A diver wearing double tanks could gain direct access to the *U-853's* interior amidships through a hole at the base of her conning tower. There, in several feet of silt, divers could search for memorabilia, most acting with a semblance of respect and decorum, but a few with incredible insensitivity. The *U-853* proved an excellent showcase for a growing controversy; should artifacts be taken from a shipwreck? The extremes of one side of the case argued no, never, better to leave it on the bottom of the ocean for history. This stance decidedly ignored the fact that the vast majority of the east coast shipwrecks were found and identified by divers looking for just these artifacts. Without the lure of artifacts, the deteriorating hulls would likely lie dormant and unseen forever. With each shipwreck disintegrating more rapidly in the corrosive saltwater each passing year, was it better to lose the unique artifacts forever, or to recover them for display above water?

The *U-853* illustrated the lunacy of the opposite extreme, the "take whatever can be carted away" philosophy. On one memorable occasion, a particularly callous diver came across a human skull inside the wreck, one of the submariners doomed to his final resting-place, or so it appeared. The diver brought the remains to the surface, proudly displayed his trophy, and stated his intentions that he wanted to fashion it into a drinking mug: maybe illegal, certainly bad taste, and definitely insensitively stupid. The charter captain promptly told the diver there was no way he would be allowed to follow through with his plan, and the skull was returned to where it had rested for thirty-seven years.

I found myself at the crux of the dilemma after recovering a piece of the *U-853* from inside her control room in 1983. With zero visibility, searching blindly by feel in the ubiquitous silt, I felt a large, breadbasket sized object and took it outside the sub and to the surface where I showed it to *Wahoo* passengers and crew. Hank Keatts, the mild mannered, unflappable southern gentleman, was a close friend of mine. In addition to being an oceanography professor, Hank had authored numerous books and articles on historic shipwrecks and diving. I held the artifact up for Hank's appraisal. It was a small air canister shrouded in leather with a confused tangle of deteriorated springs and rubber inner workings. Hank turned the artifact over, thought for moment, and made his determination: the twisted mess was a "tauchretter," a primitive escape lung; the absolute last resort by a desperate submariner to snatch a few breaths of air while attempting to claw his way to the surface after escaping from a doomed U-boat. None of the *U-853*'s crew had the opportunity to use their tauchretters. Perhaps they were trapped in the rapid flooding, or maybe the crew were all killed in the depth charge concussions, but in any event no one managed to escape.

The tauchretter, however, must have been very close at hand. Whoever had grasped it must have seen it as a last, slim hope of survival during World War Two's waning hours. All hands, forty-eight men, died on the German submarine. The tauchretter was unsettling to view, carefully displayed in my workroom, out of sight of my living room "trophies." The tauchretter evoked images of a doomed *U-853* crewman holding the small canister in a final, futile hope, before dying in isolated terror more than a hundred feet from the surface. I never displayed the escape lung prominently, but I could not dispose of it either. I had taken the thing from the wreck, and although I wished that I never had, it was my responsibility now. I couldn't think of any respectful way to get rid of it, and by the time the obvious answer came to me it had deteriorated so badly that it there was no way to tell exactly where or what it was amidst the other work room artifacts. I should have sent it back years earlier to one of my New York friends and had them return it to the German U-boat.

It was easy to get up in arms about the extremes of both positions of the shipwreck/artifact controversy. It was much more difficult to come to a rational decision in the vast, gray middle ground. Eventually, the issue would be partially settled by the Federal Government, yet it would never be truly resolved.

The *San Diego*, the *Oregon*, and the *U-853* were each intriguing in their own right. Still, there was one ship that held the fascination of the world's wreck divers in a way no other could: the *Andrea Doria*. She was huge, a nearly intact modern luxury liner that a diver could relate to easily and intuitively. She was a challenge; accessible, but just barely. Only the most experienced and competent sport divers could hope to visit her in a semblance of safety. She was historic; christened in honor of Admiral Andrea Doria, the father of the city-state Genoa, the Italia liner sought to continue the proud seafaring heritage of her namesake. But in contrast to the overwhelming success of Admiral Doria, the ship that bore his name would be forever associated with one of the most dramatic sea disasters of all time. The fact that the tragedy was followed by perhaps the world's most successful ocean rescue would enliven her history, but did nothing to remove the undeserved tarnish that would stain the illustrious Admiral's name for future generations. The *Andrea Doria's* Genoese Captain Piero Calamai would carry a more pronounced mark of shame forever. Furthermore, the *Andrea Doria* was a mystery. How could she have collided with another vessel, only one hundred miles from the mainland in well-steamed waters, with all of man's modern technology of the 1950s at her disposal, including radar?

The *Andrea Doria* was the Everest of diving, attainable with the caveat of perpetual risk and potential reward. The souvenirs from her depths were not just memories, photographs, or the ability to say that you had touched her like a mountaineer conquers a peak. A diver could return with tangible evidence that the *Andrea Doria* had been visited, artifacts beautiful in their own right regardless of the ship's history and man's drive for challenge. Diving the *Andrea Doria* inspired one's soul more than any other wreck, with thrill, discovery, accomplishment, and loss. The *Andrea Doria* was, in every respect, the pinnacle of wreck diving.

In September of 2000, I set out to return to the *Andrea Doria* and revive the thrill of exploration from seventeen years earlier.

CHAPTER FIVE

Poster Child for a Dive Accident

September 2000, Whidbey Island, Washington State

Western Washington waters are an ideal setting for deep dive training with virtually unlimited depths available close to shore, and it is not uncommon for the bottom to plunge to over 200 feet within shouting distance of a rocky cliff. The chilled waters are brimming with life - which makes for interesting sightseeing - but the living particles also filter out sun light, particularly during the spring algae bloom. Whidbey Island is at the edge of Western Washington's "banana belt," and while the surrounding mountains box Seattle in rain only an hour's drive to the south, the open, cold water mass from the Straits of Juan de Fuca provide the area with a rain shadow and significantly more days of sun than Seattle. It still rains a lot, however, and underwater lighting suffers from the overcast skies. To top it all off, river outflow thoroughly mixed by strong currents can suspend a thick layer of sediment in the water, hampering underwater visibility even further.

It looked as though we might avoid the rain for the day's dive, but not the clouds, and the tide had dropped twelve feet in the last several hours pretty much guaranteeing that the river outflow would extend into the San Juan Islands. I had not ventured beyond sport diving depth limits in over three years, and preparation for the following summer's *Andrea Doria* dives would have to start quickly; in my mind, the low tide simply meant that it would be better training.

Piloting commercial aircraft made the dive planning process more complex. Pressure continues to decrease with higher altitudes, and sufficient nitrogen "off-gassing" time must be allowed between diving and flying to avoid experiencing the bends airborne. For deep dives requiring decompression, twenty-four hours between the two activities was generally deemed adequate, but in my opinion this was not acceptable to a working pilot; if the aircraft lost pressurization and I was to get "bent" while at the controls, it would be far more serious than if a passenger in the back suffered the same condition. I set forty-eight hours as the minimum time interval between deep diving and flying. This additional buffer free of diving or flying, coupled with the constraints of tide and current, would make it difficult to train during optimum conditions. This made for more challenging training, but the *Andrea Doria* rarely presented herself in "optimum" conditions either, so I figured that was a good thing.

We set out on Chris Burgess' Uniflite for the thirty minute motor to James Island State Park at the western edge of the Rosario Straits. We were familiar with the spot from spear fishing Ling Cod, and we knew that the bottom dropped off quickly on the island's northwest point. Today would be a trial run in many respects: I had not gone diving in two months, was outfitted with a new dive computer, gauges, hood, gloves, and fins and intended to go deeper than usual. Watching the rolling island with towering trees get closer under the gray overcast, a disturbing thought crossed my mind; new gear, away from the water for a while, and going deep - I was a poster child for a diving accident.

I was anxious to start training though. Having called off dive attempts the two previous weeks for boat and dive equipment problems (hence the new gear), I needed to know if my grand plans for the *Andrea Doria* were fantasy based, or if I still felt reasonably comfortable at depth. My plan was to take it slowly, very slowly, with no greater ambition than to analyze my mental state at a minimum depth of 150 feet. The first sign of narcosis

POSTER CHILD FOR A DIVE ACCIDENT

would be my signal to turn, follow the comfort of the rising bottom terrain, and slowly ascend until reaching James Island.

Chris put the Uniflite in neutral and watched with an amused look while I donned wool socks, long underwear, and a bulky one-piece fleece coverall; maybe he saw a poster child as well, but he was used to my eccentricities and simply shook his head. I pulled on the thin, laminated synthetic shell of the dry suit itself, pausing momentarily to hold my breath before pushing my head through the restrictive rubber neck seal. Chris pulled the zipper across the back of my shoulders, leaving only my head and hands exposed to the elements. My neoprene hood and gloves would have to wait until the double tanks were secured to my back.

An accepted method of diving deep wrecks in the early 1980s was with twin tanks connected only by steel bands, each supporting a completely independent regulator. By alternating regulators every thousand pounds of tank pressure, a diver could ensure that in the event of any single system failure there would be enough air remaining in the redundant second tank to at least partially decompress and make it safely to the surface. The other common way to operate was to utilize a single manifold between the two tanks, effectively combining them into one unit with access to the air through one regulator. This alleviated the necessity of switching regulators underwater, but required the carrying of a small, emergency pony bottle with its own regulator to act as a reserve air supply if the primary regulator or tanks malfunctioned.

There were advantages and disadvantages to each method, but both setups provided redundancy through two completely independent air supplies. The average scuba tank is filled to 3,000 pounds per square inch (psi) of pressure. Most car tires are filled to about thirty psi. This tremendous pressure is reduced by the regulator's first stage, the metal housing that actually attaches to the tank valve. The air is then fine-tuned to the surrounding water pressure by the second stage, the plastic or metal housing that holds the diver's mouthpiece. It would be very unusual for a low-pressure hose operating at less than 200 psi to rupture. It is just as rare for a regulator to fail completely and without warning, although this does happen.

The weakest links in the set up are the high pressure, rubber "O" ring that seals the 3,000 psi at the tank valve and a second "O" ring that seals the tank valve to the regulator. To have one of these fail is extremely uncommon, but if it happens most of the air could be gone from the tank in less than a minute. The hose that runs to the diver's pressure gauge is another vulnerable point, but the diameter of the port leaving the regulator's first stage for the pressure gauge is specifically designed to be the size of a pin head to slow the volume of escaping air. Three months earlier I had been diving with Chris when his emergency pony bottle's rubber o-ring blew out. Although the sound of the underwater blast got my attention instantly, the backup cylinder was still almost empty by the time I swam the ten feet to Chris and shut it off.

Regulators can also "free-flow" from the surrounding water pressure when removed from a diver's mouth. Constant pressure is put on the regulator's purge button, which is normally used by a diver to expel water from the mouthpiece housing. A "free flow" releases air from the regulator uncontrollably, and while more time is available to a diver if this happens, it can still pose a significant problem.

I ran into the problem once on the *Andrea Doria* after freeing the hook and sending it to the surface with a lift bag. The *Wahoo* had grapneled into the port-side bow crane, one of two huge lifts used to provision the *Andrea Doria* pier-side. The cranes stuck out horizontally in the *Andrea Doria's* post-sinking orientation on her side. An earlier diver had shackled off a more secure anchor line and chain closer to Gimbel's Foyer Deck hole, and it was my job to recover the grapnel hook for future use. After reaching the bow crane, I filled a lift bag from my regulator and sent the grapnel to the surface, only to realize with alarm that the regulator would not stop free-flowing. I was breathing from the opposite tank on my back and there was no imminent danger, but a continued free-flow would require me to abort the dive and start decompressing immediately.

POSTER CHILD FOR A DIVE ACCIDENT

Fortunately a nearby diver, John Moyer, was able to reach the tank valve and momentarily shut off the air to the free flowing regulator. This allowed the stuck diaphragm to reset before turning the air back on. I still had to cut my dive short, but only by five minutes. What if John had inadvertently turned off the wrong tank? It should have been immediately apparent that it was the wrong tank when the free flow did not stop, but even a few seconds without a functioning regulator at 200 feet deep can be disastrous. This was one reason why on very deep dives I used to carry a small pony bottle as well, giving me a third option and a little extra air. Some of the folks who frequented the *Wahoo* found this to be overkill, but it definitely made me more comfortable. Interestingly, most technical divers in the year 2000 would find this set up, even with the extra pony bottle, woefully inadequate for a wreck dive in excess of 200 feet deep.

The bottom line is that when diving to any significant depth it's imperative that some sort of back up air source is available. Why not rely on your dive buddy as a backup? The reality is that it's exceedingly difficult to maintain close enough contact on certain dives for a buddy to be effective help. It would certainly be impossible in the silty mess of the *Andrea Doria* Dish Hole. Deep wreck diving required a great deal of self-sufficiency. Contrary to the general understanding of diving, not everyone likes to dive with a buddy anyway, either due to the purpose of the dive or the lack of a buddy whose capabilities and desires matched your own. It is exceedingly difficult to dive with a buddy in limited visibility while spear fishing. The majority of each diver's time is spent keeping track of their buddy while pointing their spear gun in a safe direction. It just doesn't work well. Similarly, when penetrating certain wrecks, in severely limited visibility with or without a penetration line, a diver might feel safer going it alone.

My strategy to prepare for the 2001 *Andrea Doria* dives acknowledged that there was no one available to train deep with me. Solo diving was viewed as a taboo subject for many years and only relatively recently have divers advocated its benefits in public forums. It is definitely not for everyone, but if done with planning, the proper equipment, and experience, solo diving is reasonably safe. Safe diving in general requires a degree of self-sufficiency. A buddy is a great additional safety factor, but on a deep wreck every diver is ultimately on their own.

The key to making solo diving reasonably safe lies in the redundancies of all critical dive gear. My backups included two completely separate tanks with independent regulators and three lights. If my buoyancy compensator somehow failed, positive buoyancy could be achieved by inflating my dry suit, and vice versa. My gear also included two separate computers to judge depth, time, and required decompression. Small, submersible dive computers were an innovation that was introduced after my *Andrea Doria* dives in the 1980s. There were two buckles on my weight belt to avoid inadvertently flipping a hasp, losing neutral buoyancy, and shooting to the surface for an almost guaranteed case of the bends or an embolism.

There were three knives on different locations of my gear to cut free of nets and monofilament fishing line. The reason for three came from a friend's experience years earlier. Chris Dillon had been swimming up the side of a New York wreck at 160 feet deep when a lost fishing net billowing in the current completely enshrouded him from above. There was no ready escape, the net was hopelessly tangled and caught in a dozen separate portions of his tank valves, regulators, and various pieces of accessory gear. Dillon took out a knife and began to cut, but soon realized that his limited maneuverability did not allow him to reach the netting behind his head. He removed his tanks and turned around while balancing them awkwardly with one hand. In his juggle to cut, hold onto his tanks, and not get entangled further, he dropped the knife and watched helplessly as it fluttered to the bottom just out of reach.

No problem, Dillon took out his second knife and again set to work cutting. Then Dillon dropped the second knife. There were no other divers in view and Dillon was trapped in the net completely alone. Without his knife Dillon was a dead man, living on borrowed time until his tanks ran dry. He swam down toward the sand straining against the fishing net's pull, stretched to his finger's limits and barely grasped the second dropped knife. Ten anxious minutes later he had cut himself free, donned his tanks, and was on his way to the surface after the tether of a lengthy decompression.

POSTER CHILD FOR A DIVE ACCIDENT

Dillon didn't wait for Christmas to buy a third knife. Diving would never be risk free, but a concerted effort to provide for redundant equipment, planning, and a healthy respect for the alien environment could make it a reasonably managed risk.

I snapped together the fittings on my buoyancy compensator and Chris went back to the helm to steer us toward the rocky shore of James Island. I waited for the stern to swing toward the rocks, stepped over the side, and slowly finned on the surface to shore while Chris shifted the boat to reverse and moved away. With regulator in mouth, I pulled down on the inflation/deflation control on my buoyancy compensator, and the back mounted bladder expelled its air allowing the thirty-five pound weight belt to slowly drag me underwater.

I equalized the increasing pressure in my ears with near continuous jaw jutting and valsalvas during the twenty seconds of descent before seeing the outline of the bottom. Identical to the uncomfortable pressure changes in one's ears created during an airplane ride, water pressure can be easily equalized by trying to exhale through the nose while pinching it closed. This is called the "valsalva" technique. Moving my jaw back and forth also worked for me. The rock wall dropped off abruptly, and I pumped air into my dry suit to relieve the pressure from the surrounding water that squeezed my body in the folds of the suit. I settled on my knees at fifty-five feet, flicked on the light at the lanyard around my wrist, and continued down.

The wall was rugged, with car-sized boulders sitting in a confused jumble, and smaller, refrigerator sized rocks filling in the gaps. Barely able to see ten feet in front of me near the surface, at eighty feet all the sunlight was completely blocked. A juvenile, eight pound lingcod swam by my light, and several small rockfish scurried away at the first hint of my presence. I continued down into the absolute blackness, pausing at one hundred feet to turn on one of the backup lights snapped to my waist, not wanting to have to fumble for it in the dark in the event the primary one failed.

At 125 feet I stopped again and took stock. The terrain had flattened considerably and boulders were no longer scattered about, but had been replaced by small rocks on the mud bottom, none larger than a foot across. I turned over the console at the end of my pressure gauge and looked at the compass - 340 degrees would steer me directly out into the wasteland of mud. After pumping more air into my dry suit and buoyancy compensator I kicked out along the gradual downward slope.

I swam on and on, frequently glancing at my computers, surprised that it was taking so long to get deeper. The distance we had marked out on the surface with the Uniflite's depth sounder had been deceiving; this was turning into a serious swim. At 140 feet deep, I halted my slow progress, checked the pressures in my tanks, switched regulators to the full, right cylinder on my back, and pressed on.

There was nothing in front of me, no fish, not even a crab interrupted the beam of my light. I became gradually reacquainted with the disturbing solitude of a deep, low-visibility dive; the absence of all dimensional perspective from the senses, a reliance on a trusting self-control from within, devoid of the reassurance of light, sound, or even touch. It was an absolute isolation, a feeling not experienced by me in years, what seemed a lifetime. The darkness and nearly flat surroundings were extremely disorienting, and without the reference to magnetic north it would be virtually impossible to distinguish between up and down slope, between a return to the light and an eternal journey to the depths. My breathing remained steady, normal sounding with none of the pinging my memory associated with narcosis. The digital computer depth readout passed 160 feet.

The indistinguishable bottom features did not change between 125 feet and where I halted my forward progress at 166 feet to take stock of my mental condition. I felt fine, slightly disoriented when not hawking my compass course, but that could be attributed to the blackness, the nearly flat bottom, and a billowing layer of fine silt after my fin inadvertently brushed the bottom. There was no significant narcosis, at least none that was clearly recognizable, but that was the problem with narcosis - it could sneak up suddenly and hit intensely.

I turned around and began the long swim back on the reciprocal heading of 160 degrees, thankful in the silent underwater that the return course

POSTER CHILD FOR A DIVE ACCIDENT

had been firmly instilled in my mind during the descent; figuring the wrong heading with such a gradual bottom slope could mean swimming away from shore, which could quickly evolve into a dangerous situation for an inexperienced diver. If the confused diver accidentally went deeper, their narcosis would increase, adding to the confusion, and start a deadly cycle of greater and greater disorientation. It has been speculated that this exact situation had killed several divers in these waters in recent years. Finally, at 125 feet, my light cut the shadowed outline of the reassuring boulders I had left fifteen minutes earlier.

I played it conservatively during my decompression "hang." One minute at twenty feet and seven minutes at ten feet were required. I followed the lively scene along the kelp bed; the interplay of fish, anemones, and crabs, and stayed down for an extra ten minutes after the most conservative computer had cleared me for a safe ascent to the surface. Bored with watching the various sea critters, I practiced raising my mask and breathing with face bared in the forty-eight degree water, checked the accessibility of each piece of dive gear with eyes closed, and worked hard at re-instilling a natural comfort level. Chris had been following my bubbles and he maneuvered the boat to my side soon after surfacing. It was a good dive; it hadn't been the depth of the *Andrea Doria*, but it was a solid start.

It was not until the following week that I managed to get into the water again to ensure that my plan was solid and I really wanted to head back east to dive the *Andrea Doria*. To make certain, I needed to get deeper to check if my comfort and resolve dissipated under the weight of over seven atmospheres of pressure.

Ron Martin was barely able to navigate the SeaRay through the dense fog under the Deception Pass Bridge on our way to the dive site. It was appropriate; the chilled Washington fog offered the same sense of foreboding and excitement felt seventeen years earlier anchored over the *Andrea Doria*. Fog was common off the coast of New England. It was caused by the collision of the cold Labrador Current and the warm Gulf Stream, and it played a big part in the sinking of the *Andrea Doria* in 1956. We skirted the coast for the short trip to the dive site with my attention riveted to the depth sounder. Nearing the cliff at Sara's Head on Fidalgo Island, the digital depth sounder readout cycled past two hundred feet and I got ready.

I considered the dive successful after exceeding two hundred feet in the pitch-black waters and assessing my mental state. I only stayed for a minute, but it all felt good; there was not even the peculiar buzzing in my ears experienced so long ago on the *Doria*. The depth put me on edge, hopefully making me a bit overcautious, definitely slightly uncomfortable in the dark emptiness. I came away with a renewed respect for the depth.

The two dives were reassuring in that I could still handle the *Andrea Doria's* depth, at least for a very short time. But the *Andrea Doria* had myriad hazards, and depth was only one. Still, it was a good start. I went home and began to lay out the plans that would take me to my goal next summer; to revisit the feel of the *Andrea Doria* under my gloved hand.

CHAPTER SIX

"One for all, all for one, and three for a quarter."

Wreck divers almost universally accept the brickwork surrounding a fireplace as an appropriate final resting spot for artifact treasures retrieved from sunken vessels. The traditional look and feel of brick lends a sense of history to the display, a shrine-like alcove that is automatically set apart from the mundane furnishings of a modern dwelling. One look across the downstairs family room at my home on Whidbey Island and it was obvious that I shared this view. The mantelpiece was decorated with various hunks of glass, brass, and lead that had each spent dozens to over one hundred years underwater. Among the array of knick-knacks were a porthole, silverware, and an elaborate five-pound lead wall ornament from the wreck of the *Oregon*, the steam-sail that sank in 1886. Vintage rifle cartridges from 1916, china, and a porthole from the U.S.S. *San Diego* also surrounded the firebox.

There was one item that stood in stark contrast to the others in size and shape; a rectangular window, with glass intact, leaning its forty-two pound weight against the brick face with unassuming simplicity. The polished brass, sparkling glass, and the single, eminently functional pull handle spoke of a design with no hidden features, no story to tell. Within the same eyeshot lay a clue to a different tale; a picture of me at twenty-two with a tired grin of triumph on my face, tousled hair still wet from the dive, shivering after an hour in the fifty-five degree water and holding a vague spirit of the same window. Anemones and rust from a steel underwater neighbor covered the Promenade Deck window, one of the glass escape routes that the *Andrea Doria's* passengers had looked through on the ship's high, port side while desperately searching for inbound rescue vessels.

One glimpse of the photograph and it was difficult to view the restored window on the hearth in the same light. What had the tired eyes that peered through her that dark, foggy night in 1956 seen? I remembered trying to look through the window after bringing it to the surface in 1984, knowing full well that no one else had been able to do so in twenty-eight years. The window had been pinned under a heavy metal frame, flush against the outboard bulkhead that circled the Promenade Deck's inner compartments, now the "floor" of the *Andrea Doria* as she lay on her side. The distorted image seen through the obscured glass on the deck of the *Wahoo* must have been similar to the one viewed that night in 1956: the window was still in the exact spot in the ocean where the *Andrea Doria* sank, where the people standing behind the window had been rescued. Instead of the confused panic that clouded the vision of the passengers, my view was clouded by the passage of time, by the natural reclamation by the sea of all things man made.

In the background of the photograph kneeled Gary Gilligan, carefully putting something away, perhaps an artifact, maybe a piece of his dive gear. Gary was the fourth in our core group of divers. Gary was a civil engineer and a general contractor from Connecticut, but if you asked him what he did for a living he would respond, "I mostly bang nails." His California-like, perpetually tanned, dirty blond appearance made him look a lot younger than his actual thirty-one years of age. His thin mustache and serious demeanor, at least to the casual observer, did not make him look any older.

"ONE FOR ALL, ALL FOR ONE, AND THREE FOR A QUARTER."

THE AUTHOR WITH PROMENADE DECK WINDOW. FROM THE AUTHOR'S COLLECTION; PHOTOGRAPHER UNKNOWN (1984).

Like Craig, the tattooed and bearded ex-biker, Don the Hook Nosed Bastard welder, and me, the dumb college kid who couldn't even rebuild an engine, Gary didn't take himself too seriously. Gary's easy going self-reliance was epitomized in 1986 when he showed up at the *Wahoo* one morning bald as a cue ball. Craig and Hook had not seen him in several months, and I had left New York to join the Navy the previous year. Hook relayed to me later everyone's curiosity regarding his new hair style. Gary explained matter-of-factly something about an inoperable cancer being wrapped around his aorta, but the hell with it, he was taking a break from his regimen of chemo to go diving. Maybe Gary looked at it simply as another competitive challenge flung down from above, but what else was he going to do about it, lie down and die? Hardly - Gary's cancer went into long term remission.

SETTING THE HOOK

We considered ourselves decent divers, but more importantly as guys with a sense of easy going satirical fun and loyalty. There was little in our lives worth getting all "bent" out of shape over. Our light hearted - and headed - Musketeer motto was, "One for all, all for one, and three for a quarter." The china retrieved on the *Andrea Doria* expeditions was equally divided among the four of us, regardless of who found each piece. A good thing for me, too; these guys were better at bringing up the dishes than me.

LEFT TO RIGHT: DON ("HOOK") SCHNELL, CRAIG STEINMETZ, PETER HUNT, AND GARY GILLIGAN. FROM THE AUTHOR'S COLLECTION; PHOTOGRAPHER UNKNOWN (1983).

Craig, Gary, and I jumped into the water in the late afternoon sun of July, 1984 with the express purpose of getting me a Promenade Deck window. The rest of our group had each managed to bring one to the surface, Craig a single unit like the current object of our interest, and Gary and Hook each had one from a giant fixed window frame that the two and

"ONE FOR ALL, ALL FOR ONE, AND THREE FOR A QUARTER."

Steve had lifted off the wreck the day prior. Craig had seen "my window" on an earlier dive, but due to its location had been unable to raise it to the surface. This second dive of the day would be a short one; we had gotten out of the water from our first dive at 8:00 that morning, and the nitrogen level was still reasonably high in our bloodstreams nine and a half hours later. Craig and I plotted strategy, while Gary listened in case we needed help. He intended to loosely tag along and survey the area in the general vicinity of our dive. Hook had made a slightly longer dive in the morning and decided to skip the afternoon's adventure.

HOOK, CRAIG, AND STEVE (LEFT TO RIGHT) WITH THE RECOVERED WINDOW FRAME. FROM THE AUTHOR'S COLLECTION; PHOTOGRAPHER UNKNOWN (1984).

We were still discussing our plan as we propped our double tanks on the *Wahoo's* doghouse and finished gearing up. It was 5:30 in the afternoon, and we planned on no more than fifteen minutes on the wreck. Still, if we were to have a problem and surface away from the anchor line, there would not be much daylight remaining for Steve to conduct a search in the open

SETTING THE HOOK

ocean with the *Wahoo*. A glance over my shoulder showed me that Gary was ready. I gave a nod to Craig and we stepped over to the cut out in the *Wahoo's* railing.

Hook helped maneuver Craig and his bulky gear toward the edge of the boat and then handed him a line that ran to the bow. Craig shifted the rope to his right hand and with a mid-air twist jumped out into the current. He hit the surface, submerged a few feet, and started pulling hand-over-hand toward the bow where the rope was shackled, allowing it to move freely up and down underwater and away from the increased drag of the surface. Craig's bubbles shot out behind him, yanked away by the swift current, as he steadily inched toward the bow trying to expend as little energy as possible. Once Craig was clear of the water below, I jumped in and reached for the guide line to the bow, quickly finning and pulling forward so Gary could do the same behind me. The spare tank and regulator dragging behind in my left hand slowed me down, but while I may not have known how to do a valve job or adjust a car's timing, I was still the strongest swimmer in our group. We brought the extra tank along to fill lift bags without depleting our breathing air.

The visibility during the descent down the anchor line was similar to the morning's dive, and while the sun had traversed the sky during our interval on the surface, the angle on the horizon was the same, but 180 degrees out. The only difference was that with each minute underwater the natural light reaching the *Andrea Doria* would lessen with the setting sun. The timing would be exceedingly close, but we were determined to get that window; it was a long, expensive trip out to the *Andrea Doria* and every dive counted.

The *Wahoo Andrea Doria* expeditions usually lasted three days. The *Wahoo* would leave her passenger loading point at the Montauk Point dock at midnight and arrive over the wreck mid-morning the next day. After the hook was set, everyone would plan on making a single dive. Two dives could be made on the second day, but the surface interval between the pair was not long enough to vent off all the residual nitrogen from the morning's dive. The second dive of the day was typically slightly shorter, and shallower, but

"ONE FOR ALL, ALL FOR ONE, AND THREE FOR A QUARTER."

it still required just as lengthy a decompression. On the third day of the trip, enough time "off-gassing" would have elapsed by mid-morning that the divers could start out fresh for planning purposes, without any residual nitrogen penalty time to consider. If all went well, each diver would make four excursions to the wreck for a total of about one hour and twenty minutes actually exploring the *Andrea Doria*. The *Wahoo* would weigh anchor around noon of the third day.

Craig and I had our work cut out for us. We were limited to fifteen minutes bottom time at a maximum depth of 190 feet. We would need to swim seventy-five feet from the anchor line along the *Doria's* hull just to get to where Craig had seen the window, with only minutes remaining to put our rehearsed plan into action. Craig didn't turn his head to check my progress, but instead rapidly kicked down the anchor line to the wreck 170 feet below. I stayed close on his fins, careful not to loosen my grasp on the spare tank. The tank was tied off to a line and snapped to my buoyancy compensator, but if I were to release the tank valve, the cylinder would bounce into my legs as I kicked.

Craig did not wait for me to settle on the *Andrea Doria's* hull before he struck out for the Promenade Deck. A quick glance at my bottom timer showed two minutes. I pumped air into my dry suit to equalize the pressure and raced after him. Feeling for my light with my free right hand, I unhooked and clicked it on without pausing. The visibility was pretty good, maybe forty feet, but it was getting dark. Judging underwater visibility is never clear-cut; one person's estimation of ten feet could easily be another's thirty. Part of the problem was individual interpretation. Varying light levels and the water clarity degradation due to sediment and plankton influenced visibility as well, but ultimately the problem was that there existed no distinctly definable topography underwater to judge as one would an eye chart. A diver might be able to see lighted objects five feet away quite clearly, but only murky, green outlines for the next twenty feet. A dark and murky forty feet could be less comfortable than a bright and clear twenty.

Craig hesitated ahead of me, adjusted his body position until he was vertical, and dropped down into the recessed Promenade Deck. I waited

SETTING THE HOOK

just long enough to avoid descending directly on top of him before doing the same. Gary was visible out the corner of my mask about twenty feet away, looking down into the recessed passageway.

By the time my knees settled onto the bulkhead of the Promenade Deck, Craig had already unsnapped the 250 pound lift bag from his side and had it unfolded. I crawled closer and glanced at my instrument console: 185 feet deep and five minutes underwater. Pulling the spare tank and regulator around from my left side, I checked that the air was on and waited for Craig to rig the lift bag. Fifteen seconds later he had the rope attached to the bottom of the heavy-duty balloon wrapped around a corner piece of the window frame and snapped secure. I placed the spare regulator's mouthpiece under the narrow opening of the lift bag, pressed the purge button on the regulator's second stage, and shot air forcefully into the narrow neck at the bottom of the lift bag until it had inflated to full capacity. The heavy metal window frame didn't budge, but stayed pinned over the solitary Promenade Deck window that had fallen free during the *Doria's* death throes twenty-eight years earlier. We had expected that the lift bag's buoyancy would not be enough; the steel frame trapping the window was huge and probably weighed at least half a ton. I dropped the spare regulator and dug my fingers under the frame at the spot where the lift bag was attached. After getting a firm grip I hunched my upper body, stiffened my torso and pushed upward with thighs against my knees planted on the bulkhead. I strained and watched as Craig tried to wiggle the edge of the pinned window.

The window moved. I relaxed for a moment, took a deep breath, and pulled again. Craig's determined back and forth motion broke the inertia and the Promenade Deck window started to slide out. After a brief rest, we each took a breath and strained one last time. Craig pulled the window free.

There was barely enough visibility in the stirred up silt for a quick high-five before I swam to the now free Promenade Deck window while unsnapping a second lift bag from my waist. I wrapped the lift bag's lanyard around the window handle, tied it in a square knot, and then snapped

"ONE FOR ALL, ALL FOR ONE, AND THREE FOR A QUARTER."

it to itself for good measure. I reached for the spare tank at my side and began to fill the second lift bag, while Craig pulled the dump cord to deflate the one used to help lift the frame. We finished our tasks simultaneously, and I guided the second lift bag and attached window out of the recessed enclosure of the Promenade Deck, with Craig swimming close behind as he wrapped up his lift bag and snapped it to his waist.

Once clear of the crossbeams and even with the *Doria's* hull, I pumped just a touch more air into the second lift bag and reluctantly let the window go. We rested motionless for about five seconds, following the window's ascent with our eyes. After ensuring that it was heading toward the surface, we began our swim back to the anchor line. As a lift bag rises, the air inside expands and accelerates the lift bag in its upward travel. Occasionally, when the bag broaches the surface it can be shot clear out of the water and deflate, allowing whatever it was carrying to drag it back down to the depths, usually far from the wreck and impossible to find.

To avoid losing a prize, a line should be attached to the lift bag and tied to the wreck or the anchor line. The line can be pulled up with the artifact this way on a subsequent dive even if the lift bag dumps at the surface. We had no time for this luxury of insurance today - I looked at my bottom timer while the lift bag went out of sight: thirteen minutes. Two minutes to get back to the anchor line and start our ascent. Gary was waiting for us at the tie in to the wreck and we followed him up the anchor line, showering him from below in our combined bubbles.

The forty minute decompression was a mixture of apprehension, relief, and unbelievable excitement. Had the window made it to the surface? Had the crew topside been able to retrieve it with the inflatable chase boat? We would have to wait to surface before learning that the window had been picked up by the chase boat without incident. The thrill slowly dissipated during the long decompression hang, giving us the opportunity to think more and more about what that window meant, how for the first time in twenty-eight years humans would look through it again. I wondered what the last person to direct their gaze through it had felt, what had happened to him or her?

SETTING THE HOOK

The general circumstances of the *Andrea Doria's* sinking were well known, but without tangible detail the more human side of the story had been difficult to fully comprehend and to visualize. Recovering that window made it all real for me. When the *Andrea Doria* collided with the *Stockholm* people had been killed, and those very decks we had just visited were the scene of unbelievable panic, bravery, and cowardice. Countless horrified souls had run, paced, and fretted past that Promenade Deck window during the eleven hours that it had taken the *Andrea Doria* to sink.

We had voluntarily exposed ourselves to great risk to look through that window, but the *Andrea Doria's* passengers had no such choice. The last to look through that window's glass must have experienced greater intensity of emotion than my excitement at bringing it to the surface, but at the other end of the spectrum of circumstance, in an out of control fear at the possibility of imminent death. I was, in however a minor and insignificant way, shaping my own fate, doing this as a choice; the *Andrea Doria's* passengers had no such option.

I thought about the owner of the silver jewelry box found outside the chapel on my first *Doria* dive. Had she survived? Probably; most of the accident's deaths occurred in the initial impact. Was she running for the lifeboats on the port side when it was dropped? Was she racing for the Promenade Deck for survival or the nearby Chapel for salvation? The *Andrea Doria* had listed sharply to the right after the collision, and the port side lifeboats were unusable. How had she gotten off, which of the six ships involved in the rescue had picked her up? Who was she?

On July 26, 1956, the *Andrea Doria's* hull was both a salvation and a curse for her passengers, keeping them safe only temporarily before the *Doria's* starboard list finally drove her into the deep. Until suitable rescue ships came near, the *Andrea Doria* was the only platform between the passengers and the chilled Atlantic beneath their feet. They were reluctant participants in a traumatic drama on the sea. I was eagerly pursuing the *Andrea Doria* seeking a thrill as great, but one that I, however tenuously and temporarily, controlled.

"ONE FOR ALL, ALL FOR ONE, AND THREE FOR A QUARTER."

THE STERLING SILVER JEWELRY BOX RECOVERED BY THE AUTHOR ON HIS FIRST *ANDREA DORIA* DIVE. FROM THE AUTHOR'S COLLECTION, PHOTOGRAPHER UNKNOWN (1983).

The port side of the Promenade Deck had been the instinctual destination of the *Andrea Doria's* 1,134 passengers after it became apparent to them that the great ship was in trouble. The Promenade Deck was identified as the muster station during the lifeboat drill conducted after leaving Genoa, Italy eight days earlier. With the exception of a single loudspeaker announcement, the drill had been conspicuously lacking in direction by the crew, but it was intuitively clear to most that the Promenade Deck was the boarding station for the lifeboats that hung from the *Andrea Doria's* davits.

An even more pressing reason for the passengers to mass along the port side Promenade Deck was simple geometry. Soon after the collision, the *Andrea Doria* listed dramatically to starboard, almost immediately past an angle of twenty degrees. It was natural to escape the rushing water and climb to the open, albeit foggy, night skies on the port side of the ship.

Andrea Doria lifeboat davits in the retracted, unused position. The lifeboats have long since broken off, most or all during the sinking. Photo courtesy of Bradley Sheard (1988).

Initially, very few passengers on the *Andrea Doria* knew for certain what had happened. Many had been asleep. The frantic theories ranged from

"ONE FOR ALL, ALL FOR ONE, AND THREE FOR A QUARTER."

a boiler explosion to a *Titanic* induced fantasy of an iceberg, a near impossibility at the ship's latitude. Still others hypothesized that the ship had hit an underwater mine or a small fishing boat. Only a few realized the truth; that the 697 foot *Andrea Doria* had been struck by a ship the size of the 525 foot *Stockholm*.

There was at least one passenger who knew for certain, although the magnitude of events must have been overwhelming to his senses. Chiropractor Thure Peterson had just fallen asleep when he was awakened by the sound of a loud impact and tearing steel. The vision before him would have been easy to discount as a nightmare; the curve of the *Stockholm's* bow passed closely and flung him through the air, knocking him unconscious. He was probably the only survivor to view the bow of the *Stockholm* while actually inside the *Andrea Doria*. He and a brave waiter, Giovanni Rovelli, would struggle for over five hours to free Peterson's wife from a pile of debris.

The passengers may have been in the dark as to their circumstances, but the senior members of the crew on the bridge of the *Andrea Doria* were painfully aware of exactly what had happened. Why it had happened was another matter. Fifty-eight year-old Captain Piero Calamai was a serious and experienced mariner who lived through the worst nightmare of any sea captain - the loss of his vessel. He had traveled the North Atlantic route one hundred times as the only Captain the *Andrea Doria* was to ever know. The *Andrea Doria* was not the fastest or the largest ship as the sun began to set on the era of transatlantic cruise steamships. Her builders had taken another route and designed the *Andrea Doria* as what was probably the most beautiful liner in existence, both inside and out.

The *Andrea Doria's* sleek lines did justice to her size. Her gradually rising superstructure culminated in a single, rounded stack. Her black hull, white topsides, and elegantly rounded stern made her appear to glide through the water. The *Andrea Doria's* sixteen lifeboats lined each side of the Boat Deck like sailors standing in review, ready to be lowered to the Promenade deck for boarding in the event of catastrophe. Not that any on board seriously entertained the thought that this might be possible. The *Titanic* had sunk only forty-four years earlier, but perhaps the intervening two World Wars clouded the collective memory enough to allow for hubris in the minds of her crew and passengers; many must have thought that the *Andrea Doria* was virtually "unsinkable."

The *Andrea Doria's* inside décor outshined even her beautiful exterior. Various Italian artists decorated the *Andrea Doria* with murals and panels made of ceramics, mirrors, mosaics, and crystal. There were four luxury suites aboard, each outfitted with every modern convenience and a distinct artistic motif. Two of the suites lined the passageway where I had found the jewelry box. There were three movie theaters and three pools. A mural painted by Salvatore Fiume covered more than 1,600 feet of the wall in the First Class Lounge. It depicted the works of Italy's proud heritage of art: sculptures and paintings by Michelangelo, Raphael, Titian, and Cellini. Also in the lounge stood a life sized bronze statue of Admiral Andrea Doria in full armor, the sixteenth century Genoese hero and ship's namesake. The forward curvature of the Promenade Deck housed a Winter Garden where an elaborate ceramic mosaic by Gambone was displayed.

In the *Andrea Doria's* cargo hold was an experimental car made by Chrysler valued at $100,000 at the time, and a socialite passenger was transporting his personal Rolls Royce. One hundred and ninety of the passengers were berthed in First Class with many luminaries gracing the roster, including the mayor of Philadelphia, movie star Ruth Roman, and Carey Grant's wife.

How was it possible for a ship as magnificent as the *Andrea Doria* to sink? Even professional mariners must have entertained the notion of a "nearly" unsinkable ship when considering the *Andrea Doria*. Built to the conform to standards of the International Conference for Safety of Life at Sea of 1948, she was not supposed to list any more than fifteen degrees in the worst of anticipated flooding conditions. On her maiden voyage in January 1953, she had encountered a vicious North Atlantic storm and caught a huge, rogue wave broadside. The *Andrea Doria* had listed a full

"ONE FOR ALL, ALL FOR ONE, AND THREE FOR A QUARTER."

twenty-eight degrees before righting herself, proving to be far more stable than most other vessels that would have capsized at less of an angle.

GLASS-TOPPED TABLE SECURED TO THE DECK OF THE FIRST CLASS COCKTAIL LOUNGE.
PHOTO COURTESY OF BRADLEY SHEARD (1988).

Over time, fate has a way of reminding man of the illusion of control; nothing man made is immune from disaster.

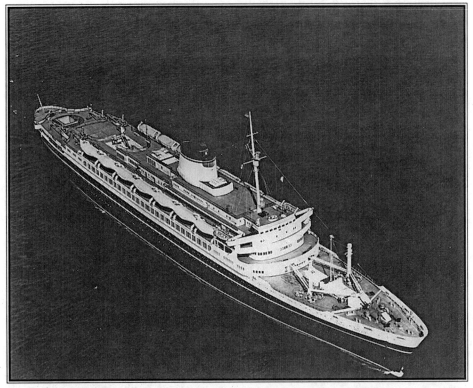

THE *ANDREA DORIA*. COURTESY OF THE ERIC SAUDER COLLECTION.

CHAPTER SEVEN

Like a high-stakes Rorschach test

July, 1956

Perhaps in a premonition of what was to come, Captain Calamai arrived unbidden to the bridge just as the fog bank came into view. Even in 1956 with radar aboard most modern vessels, fog was a recognized killer. It takes a great deal of distance to turn a ship the size of the *Andrea Doria* and the visibility required to react to an obstacle translates into lifesaving time. But concern for safety is not the only pressure on a professional mariner, although it unarguably should be the dominant influence. In a token gesture - a ship's captain is judged in part by his ability to keep a schedule - Captain Calamai reduced the *Andrea Doria's* speed from twenty-three to 21.8 knots. According to the rules of the sea, a ship should be able to stop in half the distance of the prevailing visibility, thereby theoretically allowing just enough room for two ships on opposing courses to avoid a collision by coming to a full stop without altering headings. On the night of July 25th, the thick fog had reduced visibility at times to less than one hundred yards, making it impossible for the *Andrea Doria* to meet her stopping requirement at a 21.8 knots cruise, but speeding in fog was a common maritime practice.

Without the comfort of radar onboard, would Captain Calamai have maintained such a high speed - probably not; almost certainly not in a densely traveled maritime route. Did radar instill in Captain Calamai a feeling of complacency, a misplaced reliance on technology in the stead of good seamanship? Perhaps; it certainly happens to all of us in everyday life, but is frequently overlooked because the consequences rarely extend beyond minor inconvenience. The beautifully simple computer truism, "gi-go," or "garbage-in, garbage-out," can also be applied to a person's reasoning as they work through a complicated problem. Like a high-stakes Rorschach test put into practice, operators - even those with solid training - tend to see in their display of technologically derived data what they want or expect to see. "Garbage" assumptions or interpretations, particularly those that form the basis for a complicated series of follow-on actions, are difficult to identify, much less discard as invalid.

Captain Calamai was far from alone in his reliance on technology among professional mariners, often in lieu of time honored experience. A near blind trust in technology is still seductively common practice in many highly automated fields, notably commercial aviation. Experts occasionally make beginner mistakes to varying degrees in most high risk endeavors. Only those who can translate the lessons learned from actual experience into permanent habit patterns seem to skirt the threat of turning a series of simple errors into deadly tragedy.

Other precautions were taken on the *Andrea Doria* due to the fog: more crew were sent to the engine room to work the valve wheels that controlled the ship's power plant, the prescribed foghorn signal was initiated, and a radar watch was set. And, in the event the worst possible situation developed, the *Andrea Doria's* watertight doors were closed. Watertight integrity is essential to a combatant vessel, where creature comforts are cast aside for rigid solutions of steel barriers in the event of a breached hull. This is not the case on a luxury liner, where it is difficult to deny paying passengers ready access to nearly all parts of the ship. Even with the *Andrea Doria's* watertight doors closed, seventeen stairwells provided openings in the protective transverse bulkhead of A-Deck, two levels above the ship's waterline. The sea could pour in from a hole at the waterline and travel up through these seventeen gaps in the watertight barrier if the *Andrea Doria* lay over at too great an angle; twenty degrees, to be exact. The role that the *Andrea Doria's* watertight integrity actually played in her eventual demise would remain a source of debate for twenty-five years.

There was one action that was not taken, however, that unquestionably affected the stability of the ship. The *Andrea Doria's* fuel tanks were not filled with salt water for ballast once they emptied. Captain Calamai did not want to incur the time or expense of removing the salt water and cleaning the tanks prior to refueling in New York for the return voyage to Genoa. The additional buoyancy at such a low center of gravity produced a bobbing cork effect that would greatly exaggerate a list if only one side of the ship's fuel tanks were to flood.

The precautions that Captain Calamai ordered were not unusual or insufficient from the perspective of normal operating procedures. Captain Harry Gunnar Nordensen of the Swedish-American passenger liner *Stockholm* might have taken similar action had he been aware of fog in the path of his ship. His twenty-six year old Third Officer, Johan-Ernst Carstens-Johannsen, had control of the helm, however, and from his perspective, looking at the *Stockholm's* navigation lights in the night air, there was no indication of fog ahead.

Radar was not a new technology in 1956, in fact, it was fast becoming commonplace and both the *Andrea Doria* and the *Stockholm* were operating their radars as they traversed the "Times Square" of the Atlantic in opposite directions. All ships coming to New York from Europe or leaving New York to the east had to travel this general path in order to turn outside the shoals protected by the Nantucket Light Ship. The 1948 International Convention for the Safety of Life at Sea advised eastbound traffic to steam a route twenty miles to the south, but the *Stockholm* was not a party to the agreement. Third Mate Carstens was directing the helmsman of the eastbound *Stockholm* to a point one mile south of the Nantucket Light Ship; time was money and Captain Nordensen always followed this route.

The fact that two large ships would pass so closely at night, possibly in the fog, might seem foolish to the layperson. But this was common practice, and both vessels expected to encounter opposite direction traffic in the vicinity of the Nantucket Light Ship. There were firm rules of the road for the passage of ships at sea, rules that were designed to separate approaching ships and prevent a collision. To successfully utilize the rules of the road each vessel needed to have a clear understanding of the other ship's course, preferably with a visual sighting, but otherwise through the use of radar.

To this day, radar - on the water or in the air - is not foolproof and is still subject to the interpretation and skill of the operator. But modern radar, particularly when interfaced with a computer, is vastly superior to what was available in 1956. The concept of radar is simple. A pulse of energy is emitted from a transmitter and if it strikes an object, be it land or another vessel, it returns to the radar receiver for display on the radar screen. The time in fractions of a second that it takes the energy pulse to travel is translated into distance to the radar target. If the energy pulse hits nothing in the set distance range it dissipates before it can return to the receiver, and no objects are therefore depicted on the radar display. A 1956 radar screen showed nothing more than the returned energy pulse in the form of a light "blip." It was up to the radar operator to interpret the changing positions of the blip in relation to his own ship to estimate another vessel's course and speed.

On the *Stockholm* at approximately 10:50 pm, Third Mate Carstens, the only officer on the bridge, saw a radar blip twelve miles away. Carstens needed to plot the trend of the blips on subsequent sweeps of the *Stockholm's* radar to translate the small light flashes into meaningful information. A single radar return only told Carstens where the unknown vessel was at the exact moment when the radar energy was reflected and it told him nothing of the ship's course, the singularly vital piece of information required to avoid a collision. To determine the closing ship's course, Carstens needed to mark each consecutive radar blip on a separate "plotting board" and then calculate the trend in the other ship's movement.

Carstens split his time plotting the radar blips, searching the clear (from his vantage-point west of the Nantucket Light Ship) horizon for lights, and monitoring his helmsman who had a tendency to let his course wander. He was also required to answer the phone in the event one of the lookouts on the bridge wing reported observing lights. As the *Stockholm* approached the

LIKE A HIGH-STAKES RORSCHACH TEST

Andrea Doria, twenty-six year old Third Officer Carstens, the oldest person on the bridge, was a very busy man.

The bridge of the *Andrea Doria* was flush with officers of the Italia Line. In fact, there were two licensed ship's masters present in addition to Captain Calamai: Second Officer Curzio Franchini and Third Officer Eugenio Giannini. Second Officer Franchini monitored the *Andrea Doria's* radar as the speeding liner raced past other vessels heading toward New York. When a radar blip appeared almost directly ahead of them, the rate of closure made it immediately apparent that the two vessels were on opposite courses and would pass very close to one another. Franchini did not take the time to mark each successive radar blip on a plotting board, as this was not standard procedure in the Italia Line. He instead tried to judge the trend from one blip to the next to form a mind's eye rough estimate of the phantom ship's course.

For the *Andrea Doria* and the *Stockholm* to pass one mile apart would not have been noteworthy. In fact, Captain Nordensen on the *Stockholm* had left instructions that he need not be called unless another vessel was to pass closer than the one-mile safety zone. International rules of sea mandated that ships pass left to left - much like cars in the United States - unless it would clearly require exceptional maneuvering to reach this orientation. If this were the case or if there existed any question as to the geometry of the crossing, both vessels were required to make a sufficiently significant course correction to signal their intention of making a right to right pass to the opposing ship.

Nineteen fifty-six radar was a simple, mechanical/electrical device. Interpreting the information of a radar display, even today, can be an art. On 25 July 1956, radar interpretation was the only thing separating the *Andrea Doria* and the *Stockholm* in their nearly head-on race together at a combined speed of approximately forty knots. Neither ship's crew would have sufficient visibility to accurately determine the other's course until a collision was unavoidable.

But for now, all Third Mate Carstens saw out his bridge window was clear sky, and he had no idea that the *Andrea Doria* was racing toward his ship enveloped in the leading edge of a fog bank. Had he been aware of fog, he almost certainly would have followed the *Stockholm's* standing orders and sent for Captain Nordensen to return to the bridge to assume command and lighten the workload. When Carstens plotted out the radar returns it was apparent to him that the *Stockholm* and the unknown ship approaching from the east would safely pass left to left. Captain Calamai and Second Officer Franchini on the *Andrea Doria* interpreted their radar screen very differently; they viewed the passage as one of right to right. The officers on the bridges of the two approaching ships were operating on antithetical assumptions.

The *Andrea Doria* and the *Stockholm* were set up to pass so closely that determining whether it was a right-to-right or left-to-left orientation would have been difficult, but certainly not impossible. Regardless, the two ships were directing their bows to well within a one-mile distance from one another, much closer than either Captain Calamai or Third Officer Carstens wanted. No one will probably ever know with absolute certainty if there was a single dominating factor that caused the two passenger liners to collide. There was no obvious, glaring error on either crew's part that caused the collision. As is the case with most accidents of professional mariners or seasoned aviators, it was a chain of seemingly minor circumstances, errors in interpretation, and faulty judgments that led to disaster.

Third Mate Carstens had been gradually correcting the *Stockholm* for a northward drift caused by the wind and tide. By directing the helmsmen to steer several degrees in a more southerly quadrant, Carstens was routing the *Stockholm* at a slight angle to the direct line between the Ambrose Light Ship outside New York Harbor and the Nantucket Light Ship; the *Stockholm* and the *Andrea Doria* were therefore not on exact reciprocal courses. The *Stockholm* was steering 091 degrees, the *Andrea Doria* 268 degrees. There existed a three degree crossing angle, an angle so small that it would have been difficult to detect from radar returns alone.

Carstens was task saturated. He had to divide his time at the radarscope, in the chartroom plotting his radar findings, and at the helm verifying the *Stockholm's* course. There was no "gyro repeater" to read the *Stockholm's*

LIKE A HIGH-STAKES RORSCHACH TEST

heading at the radarscope, and Carstens had to rely on a verbal relay of heading information from his helmsman, a sailor with documented difficulty maintaining a steady course. A minor discrepancy in Carstens' understanding of the *Stockholm's* actual course would produce an exponentially larger error in his plot of the two ships' relative paths. Carstens was busy, but in the clear night air he felt no discomfort or unease at the prospect of navigating his ship as the only officer on the bridge.

There may have been an additional problem. Under Carsten's workload it is possible that his radar was adjusted to a five-mile range setting versus the fifteen-mile reference that he assumed for his plotting. If this were the case, the *Andrea Doria* would have been only two miles away with her lights still concealed in a fog bank when Carstens was operating under the illusion that she was actually six miles distant. The lookout on the *Stockholm* strained his eyes intently into the infinity of black, night horizon. On the ocean at night, away from the lights of shore or other vessels, attempting to judge distance can be extremely disorienting. A dim light on a clear night might be dozens of miles away. The same light viewed in a mist, or fog, might appear at five, three, or at less than one mile away.

Finally, the *Stockholm's* lookout began to see the *Andrea Doria's* lights. Carstens went outside on the bridge wing for a better view. Still unaware that the *Stockholm* was on the edge of a thick fogbank, Carstens relied on his radar to determine the distance to the *Andrea Doria*. His interpretations of the ill-defined and what looked to be distant, lights of the *Andrea Doria* still indicated that the two ships were on a left-to-left passage. But they would pass closely.

Carstens ordered a significant turn to the right, more than twenty degrees. A turn of this amplitude would widen what he perceived as the two ship's left-to-left orientation to the minimum one-mile distance that Captain Nordensen had mandated in his standing orders. For an unknown reason, perhaps because he thought he was much further from the *Andrea Doria* than was actually the case, he did not signal his maneuver with the *Stockholm's* horn as was required.

The *Stockholm's* course change to the right would have had the desired effect of widening the gap between the two ships if the distance between them was indeed six miles. If, however, the *Stockholm's* radar range setting was inadvertently set to only five miles, Carsten's plotted course of the *Andrea Doria* would be erroneous, and the *Stockholm's* bow would be turning into the *Andrea Doria* only two miles away. The *Stockholm's* turn occurred far too close for the *Andrea Doria* to react once they discovered it on their radar, and the *Andrea Doria's* officers would have to rely on a visual sighting of the *Stockholm* to learn that she was turning into them.

On the bridge of the *Andrea Doria*, Captain Calamai interpreted his radar as indicating the ships would pass closer than one mile as well, but he presumed the passage would be right to right. At three and a half miles, he ordered a left turn of four degrees to a 264 degree heading. He reasoned the turn would widen the gap in an already established right-to-right crossing. Four degrees was not a large enough correction to signal the danger to Carstens on his radar, now less than four miles away; the *Stockholm's* radar plot would not reflect the course change until it was too late. The crossing angle between the two ships had increased to twenty-seven degrees.

The tragic irony was that the *Stockholm*, under the impression of a left-to-left pass, was under no obligation to make a dramatic course change to signal her intent - yet she did. The *Andrea Doria*, laboring under the assumption of a right-to-right passage, was burdened with broadcasting her intentions with a turn of obvious magnitude - yet she did not.

When the officers on the bridge of the *Andrea Doria* finally obtained a visual sighting of the *Stockholm's* lights, they were barely a mile apart. Ships have two white masthead lights, a lower one forward, and a higher one aft. They also illuminate a green position light on the ship's starboard side and a red one to port. The combination of lights enables a practiced observer to quickly judge the ship's course relative to their own. Third Officer Giannini strained to see a clear picture through his binoculars to determine the correct aspect of the *Stockholm's* light picture. Within seconds, at less than a mile away, with a combined closure rate of forty knots, it became evident that the *Stockholm* was turning directly into the *Andrea Doria*. The time required for a turn to take effect on a ship the size of the 697 foot *Andrea Doria* cannot be overemphasized. If the *Andrea Doria* had turned her bow into the conflict, to the right, the two ships would have collided, possibly at as great a combined speed as forty knots, but they would have met bow to bow, and the impact would have likely been absorbed in a glancing blow and the subsequent pivoting of the two ships.

Captain Calamai swallowed his horror and made a split second decision. He called for a turn hard to the left, while maintaining the *Andrea Doria's* speed to expedite the turn, in the impossible hope of avoiding a collision altogether. But a collision was inevitable by this point, and the only question remaining was the angle of impact when the two great vessels came together. By turning left instead of right Captain Calamai vainly hoped to skirt a collision; instead, he ensured a disaster.

Carstens on the *Stockholm* lost sight of the *Andrea Doria* as he left the bridge wing to answer the phone - a second lookout was also reporting the lights of the *Andrea Doria*. The ships were closer than one mile when Carstens returned his gaze to the *Andrea Doria*. Amazingly, from his view point, it appeared that the *Andrea Doria* was turning in front of the *Stockholm*. Realizing that a collision was unavoidable, Carstens ordered full reverse and a hard turn to starboard.

The chain of events that led to the collision in the next few seconds could have been broken by any of a number of actions or circumstances. If there had been no fog any confusion due to differing radar interpretations would have been resolved visually while still many miles away. If Carstens had been aware of the fog earlier, he would have summoned the Captain to take command, thus lowering the workload. Carstens could have better operated and interpreted his radar. He could have signaled his dramatic, twenty-two degree turn to the right to the *Andrea Doria* with the *Stockholm's* horn.

The *Andrea Doria* was steaming too fast for the conditions of visibility, although not to do so would have been unusual for the scheduled liners of her day. The officers on the *Andrea Doria* did not plot the course of the *Stockholm* as depicted by the radar. Captain Calamai could have ordered a much larger turn to the left at three and a half miles to better signal his intentions for a right-to-right passing. Finally, Captain Calamai could have turned right when the collision was unavoidable.

SETTING THE HOOK

At approximately 11:10 pm, on 25 July 1956, the *Stockholm's* bow - reinforced for northern ice - sliced broadside into the *Andrea Doria's* starboard side just below where her ship's officers looked down in disbelief from the bridge. Chiropractor Thure Peterson saw it with his own eyes from cabin fifty-six in the *Andrea Doria's* Upper Deck. The *Stockholm's* bow split five decks and further down into the empty deep-water fuel tanks on the *Andrea Doria's* starboard side. The *Stockholm's* bow was still inside the *Andrea Doria* as over five hundred tons of saltwater raced to fill the void left from her breached hull.

The interior of the *Andrea Doria* below the ship's wheelhouse was in shambles. The ceiling of the wide-open foyer exploded in a heap of rubble. The plate glass window of the *Andrea Doria's* gift shop miraculously survived, but trinkets scattered everywhere. The previously luxurious foyer deck had sustained the energy of a bomb blast.

The *Andrea Doria* listed almost immediately twenty degrees to starboard. The mighty ship was dying.

AFTER THE COLLISION. COURTESY OF THE ERIC SAUDER COLLECTION.

CHAPTER EIGHT

...dragged into the depths...

July, 1956

Five *Stockholm* crewmembers berthed in the bow were killed immediately when the two ships collided. Third Officer Carstens ordered the Stockholm to full reverse and watched in dazed disbelief. The *Andrea Doria's* starboard side absorbed the momentum of the impact allowing the *Stockholm's* bow to bounce back abruptly, and then continue forward again dragging along the *Andrea Doria's* side in a screeching shower of sparks. Carstens recovered his composure, ordered the *Stockholm's* watertight doors closed, and then rushed to find the ship's master, but Captain Nordensen reached the bridge before Carstens got to the door. A quick damage assessment was conducted, and the *Stockholm's* officers breathed a sigh of relief to learn that it did not appear that the *Stockholm* was in imminent danger of sinking.

Both of the *Stockholm's* anchors were unintentionally released when her crumpled bow could no longer support their weight, and three of the *Stockholm's* five dead crewmen were dragged into the depths with the anchor chains. For a few tense moments it appeared that the *Andrea Doria* and the *Stockholm* might drift back together, but Captain Nordensen was helpless to move his vessel until he could cut free the two anchor chains pinning his ship to the ocean floor. The *Andrea Doria* moved at the whim of wind and current as Captain Calamai was initially reluctant to engage the liner's port engine in fear of capsizing. Fortunately, the *Andrea Doria* drifted away from the *Stockholm*.

The *Andrea Doria* was in far worse shape than the *Stockholm*. The hole punched out by the *Stockholm's* bow caused the five empty fuel tanks on the *Andrea Doria's* starboard side to flood, and the out-of-balance difference in buoyancy twisted the ship and increased her starboard list. The *Andrea Doria's* crew attempted to counter the growing list by pumping saltwater from the ruptured fuel tanks to those on the port side, but they could not gain access through the wreckage and flooding to the proper control room to complete the task.

Captain Calamai stared dumbfounded as the *Stockholm* disengaged from his ship and drifted out of sight as the fog again enveloped the *Andrea Doria*. Finally breaking from his stupor, Captain Calamai ordered the passengers to their muster stations, but it was a call lost to many in the confusion. It would be the last announcement to the information-starved passengers for hours to come. In an attempt to avoid a panic Captain Calamai specifically ordered that no signal to abandon ship be issued, leaving hundreds of passengers hopelessly confused and with no clear direction.

The first SOS left the *Andrea Doria's* radio shack at 11:22 pm, and was received at the East Moriches Coast Guard radio station on Long Island. By the time the New York Rescue Coordination Center was notified, the *Andrea Doria* was at a twenty-five degree list, rendering all the ship's port side lifeboats useless as gravity pinned them to the ship's rail and prevented them from being lowered. The remaining eight serviceable starboard lifeboats were only enough to accommodate 1,004 of the *Andrea Doria's* passengers and crew. Fortunately, the decisive action of six nearby vessels, the *Stockholm* included, were able to prevent further tragedy.

Ships as far away as Argentina and Newfoundland heard the *Andrea Doria's* distress signal, and in short order nearby vessels from all directions were converging cautiously through the fog on the position of the SOS call. Ironically, the *Stockholm* would be the first ship to pick up *Andrea Doria* survivors, despite Captain Nordensen's concern that his vessel might not yet be clear of immediate danger. Over the course of the next nine hours,

...DRAGGED INTO THE DEPTHS...

four other ships participated in the rescue and picked up as few as one and as many as 158 *Andrea Doria* passengers and crew. But it was the massive French liner *Ile de France*, inspired by the selfless action of her Skipper, which did the lion's share of the rescue work and eventually plucked 753 *Andrea Doria* passengers and crew from life boats and the sea.

There were 1,706 passengers and crew on the *Andrea Doria* and probably an equal number of individual interpretations of the collision. Forty-six *Andrea Doria* passengers died as a result of the disaster leaving their personal perspective untold. The combined fatalities of the two ships numbered fifty-one out of a total possible of almost 2,500. Fifty-one may appear a small number when compared to what could have happened, but each death touched a network of countless other lives: children, mothers, fathers, wives, husbands and friends. Thure Peterson, the chiropractor who witnessed the *Stockholm's* bow inside the *Andrea Doria*, would certainly never look at the number fifty-one lightly; one of those numbers represented his wife. When the *Stockholm's* bow crashed into cabin fifty-six at 11:10 pm, Peterson and his wife Martha were sleeping in twin beds separated by a heavy chest of drawers. Both survived the initial impact, but where Thure Peterson was flung clear into the wreckage of the adjoining cabin, Martha was seriously injured and pinned under a mountain of rubble. Frustratingly aware that he would be unable to reach his trapped wife without help, Peterson struck out in a frantic search first for assistance, and then for morphine to ease his wife's suffering.

A Cabin Class waiter, Giovanni Rovelli, heard Peterson's shouts and responded to his urgent request for aid. Rovelli was a much smaller man than Peterson and was able to squirm through the mangled cabin to begin clearing the tangled wreckage, while Peterson rushed off again in search of morphine. The *Andrea Doria's* two doctors and five nurses had set up a makeshift aid station on the high, port side of the listing ship near the entrance to the Winter Garden. Reading the account years later, it struck me that this could not have been far from where Craig and I had recovered the Promenade Deck window. Had Peterson looked through that window, searching for answers in the foggy darkness, for ways to try and ease his wife's suffering?

SETTING THE HOOK

Peterson was eventually able to find morphine to lessen his wife's pain and that of another trapped passenger, Jane Cianferra. The cabin continued to tilt to precariously greater angles; by 2:30 am the *Andrea Doria's* list was at thirty-three degrees, and by 3:30, it was nearly forty. The two women pleaded with Peterson and Rovelli to leave them and save their own lives before the ship capsized; Peterson and Rovelli simply ignored them. The two procured an axe, wire cutters, and eventually, through tremendous effort, a lifting jack. After two hours of hard work, Jane Cianferra - her leg and arm broken - was freed and taken topside to the ship's doctors. Finally, at almost 4:30 am, Peterson and Rovelli manhandled a 150 pound jack, transferred miraculously from one of the rescue vessels, into position to free Martha Peterson. Just as the wreckage pinning Martha to the deck finally began to move, she let out a final gasp and died. Thure Peterson said a brief farewell in the darkened cabin, and then left with Rovelli for the open deck and survival.

Not all of the *Andrea Doria's* crew acted nearly as valiantly as Rovelli. Over two hundred crewmembers, mostly stewards, waiters, and kitchen help, filled lifeboats and fled the *Andrea Doria* before the passengers were evacuated. It was troubling to note that the *Stockholm* - the first to pick up survivors - carried to New York 234 of the *Andrea Doria's* 572 crew and only 311 of her passengers. These acts of cowardice surprised me initially, perhaps due to the inseparable associations in my mind between the *Andrea Doria* and diving. How could anyone connected with the liner have acted in a cowardly fashion? A diver on the *Doria* had to possess at least a bit of nerve. But, of course, the name *"Andrea Doria"* meant something entirely different to the crew and passengers in 1956; unlike the divers to follow in the ensuing decades, they had not chosen risk and adventure. The passengers and crew, the disaster's reluctant participants, demonstrated the full spectrum of human reaction to adversity from heroism to cowardice. Much of the drama following the collision of the two unfortunate liners was the result of human action and character. Still, the most incredible events of the disaster were attributable more to circumstance and chance than to human intent.

Martha Peterson's companion trapped in the rubble, Jane Cianferra, had not been traveling on the *Andrea Doria* by herself. Her husband, Camille Cianferra, was a foreign correspondent for the New York Times in Spain. This fact, coupled with the proximity of the offices of the New York Times to the disaster, must have instilled in the Times reporters a tremendous urge to scoop the *Andrea Doria* story. Times reporters waited eagerly for what they could only imagine would be a dream come true; a reporter eyewitness to a catastrophe, a first person account from a news professional who was actually in the story. It was only much later that the feelings of envy vanished when it was learned that Camille Cianferra had been killed in the collision along with the family's eight-year old daughter Joan.

But incredibly, there was more still more to the Jane Cianferra story. Jane Cianferra's second daughter, a fourteen-year old from a previous marriage to radio personality Edward Morgan, would survive in a truly amazing chain of events, illustrating how the difference between triumph and tragedy is often measured in inches. Linda Morgan had been sleeping in the bed adjacent her sister, Joan, when the *Stockholm* struck.

Linda Morgan's first recollection was crying for her mother in Spanish, her first language learned while growing up in Mexico and Spain. Coincidentally, a Spaniard in the uniform of a cleaning man heard her moans of distress and crawled through the mangled wreckage to the sound. The Spaniard was startled to find someone speaking his native language; it was the first he was aware that another Spaniard was on his ship of employ, the *Stockholm*.

While Joan was killed instantly in the neighboring bed, Linda Morgan was scooped from the inside of the *Andrea Doria* by the *Stockholm's* bow and flung upward. She landed on the deck of the *Stockholm* eighty feet aft of the collision point, behind a bulkhead that shielded her from flying metal. It was only after listening to Linda Morgan speak of the *Andrea Doria* that the *Stockholm's* crew suspected the amazing truth. Linda Morgan's phenomenal transfer from the *Andrea Doria* to the *Stockholm* would christen her forever in sea-disaster lore as the "miracle girl."

By 1:40 am the fog had cleared, and in the hours to come 1,663 *Andrea Doria* survivors and three bodies would be evacuated from the doomed ship. Forty-three souls were left behind to find their final rest on the ocean bottom. Once convinced that all his surviving charges were safely away, a despondent Captain Calamai left the sinking *Andrea Doria* at 5:30 am on July 26, 1956. After a night of horrible coincidence and mistakes, Captain Calamai would prove wrong even in his belief that he was the last to leave the stricken liner.

Robert Hudson was a merchant sailor, but not a member of the *Andrea Doria's* crew. He had sustained serious injuries to his back and right hand while working on a freighter in European waters and was boarded on the *Andrea Doria* in Gibraltar for convalescent passage back to America. His intense pain made it difficult to rest, and on this last night at sea he had taken a powerful painkiller and finally fallen asleep in his cabin hours before the collision. He did not wake until 5:10 am, six hours after the *Stockholm's* bow cut into the *Andrea Doria*.

When Hudson finally woke disoriented in total darkness, he flicked his cigarette lighter and looked around; he was being pressed by gravity against the cabin's bulkhead wall and was closer to the ceiling than the floor. The only ambient light came through the open cabin door from the emergency lamps in the passageway. Hudson immediately recognized the emergency lighting, processed their meaning, and scrambled through the gathering oil and water on the cabin floor in search of open air and escape from a ship that he intuitively knew was sinking, but had no idea where, why or how.

Fighting the pain of his injuries and the severe list of the *Andrea Doria*, Hudson reached fresh air after an hour of intense effort only to discover to his dismay that even the deck was awash in saltwater. As Hudson tried to plan his next move, a wave tumbled over the deck rail and pulled the injured merchantman into the water. Fearing that the suction from the sinking *Andrea Doria* would pull him down, Hudson reached out frantically and miraculously grabbed a cargo net that had been lowered hours earlier to assist passengers climb down to the waiting lifeboats. He hung on with the knowledge that his life depended on the strength in his fingers.

Several hundred yards away a lifeboat crew watched in horrific fascination. They did not dare come closer; the *Andrea Doria* threatened to go under any second, and the vacuum-suction from her sinking would certainly pull them down in tow. Hudson and the lifeboat crew looked at one another at a distant impasse for ninety minutes. Hudson first pleaded, and then swore like only a sailor can, and finally he began to pray. With injuries too severe to permit a swim against the suction of the slowly sinking *Andrea Doria*, he was left with no choice but to continue clinging to the cargo net in limbo. Finally, the lifeboat crew could look on no longer. They rowed hard to Hudson and pulled him aboard. Robert Hudson was the last person to leave the *Andrea Doria* at 7:30 am.

The *Andrea Doria* fought on against the sea's irresistible pull, refusing with stubborn pride to be dragged down into the depths. It was not until 10:00 am that her bow finally surrendered. Nine minutes later, the last view of the *Andrea Doria* on the ocean's surface was afforded to Harry Trask, a Boston Traveler photographer, flying in a chartered plane overhead. The name "Andrea" on the ship's stern, flanked by one enormous, motionless propeller, was the last piece of the mighty ship to grace the surface. Harry Trask's series of photographs of the *Andrea Doria*'s final minutes earned him the Pulitzer Prize for 1956. At 10:09 am, 26 July 1956, the *Andrea Doria*'s stern slipped beneath the Atlantic amid a field of floating debris torn loose from the sinking ship.

SETTING THE HOOK

COURTESY OF THE MARINERS' MUSEUM, NEWPORT NEWS, VIRGINIA.

The death toll did not end with the sinking of the *Andrea Doria*. An hour before the *Andrea Doria's* final plunge into the depths, two Coast Guard helicopters reached the *Stockholm*, where some of the most seriously injured were being treated by the ship's doctor. Early on in the disaster, at about 1:30 am, several of the *Stockholm's* lifeboats began shuttling to the *Andrea Doria's* side to evacuate survivors. The wide-spread fear that the ship would go down any second prompted desperate parents to drop their children from the open Promenade deck twenty feet above to the lifeboats below. One little girl landed in the water, but was quickly plucked from the sea by a *Stockholm* sailor before she could drown. Four-year-old Norma di Sandro was not so lucky and was released from her desperate father's grasp before the lifeboat crew could stretch out a heavy blanket to catch the child.

...DRAGGED INTO THE DEPTHS...

COURTESY OF THE MARINERS' MUSEUM, NEWPORT NEWS, VIRGINIA.

The little girl's head hit the lifeboat's gunnel, knocking her unconscious. Norma and a seriously injured crewmember from the *Stockholm* were the first to be pulled off the *Stockholm* by Coast Guard helicopters and raced to Boston for medical help.

The injured *Stockholm* crewmember died almost immediately upon reaching Boston. Two days later Norma's desperate parents finally found her. They reached her in the early morning hours at the Brighton Marine Hospital in Boston, where the little girl lay in a coma after undergoing surgery to relieve the pressure from a fractured cranium. Six hours later Norma di Sandro died quietly, never having regained consciousness.

The aftermath of the sinking and ensuing maritime trial proved to be as contentious as it was inconclusive. After six months of combative argument the Swedish and Italia Lines agreed to settle out of court. The Italia Line absorbed the loss of the thirty million dollar *Andrea Doria*. The Swedish Line accepted the cost of the one million dollar repair to the *Stockholm's* bow and the one million-dollars of lost business revenue. Both Lines pooled their personal injury liability to pay for the many third party suits.

It would be a long wait before the friends and relatives of the last *Andrea Doria* victim could begin the grieving process. Julia Greco broke her back when she fell into a lifeboat during the evacuation, and the fifty-three year old Pennsylvania woman held out for six months in the hospital enduring severe pain before becoming the final person to be claimed by the collision.

Although Julia Greco was the last to die as a result of the 1956 collision, she was not the final victim to be taken by the *Andrea Doria*.

CHAPTER NINE

"Normal" in an utterly un-normal world

July 15, 1984, fifty miles south of Nantucket Island, Massachusetts

Gary Gilligan dropped down feet first into the eerie void of the *Andrea Doria's* Foyer Deck. The relatively clear waters of the outer hull, well lit by the morning sun, were no match for the inky black of the vertical shaft and Gary quickly disappeared into the darkness. Raising the buoyancy compensator inflation/deflation hose over my head, I depressed the deflation button and slowly sank as tiny bubbles squeezed by six atmospheres of salt water streamed in a staccato of loud pings from the rubber bladder. My light beam found the emptiness where the bulkhead receded into the Dish Hole and stayed locked to it as I continued to descend into the wreck. Gary's fins disappeared down the passage, ominously vanishing as if the wreck had swallowed my dive buddy whole.

I peered down the entrance to the Dish Hole, the passageway to where the majority of my time on the *Andrea Doria* had been spent, and savored the comfort of the familiar surroundings. The light faded from the corridor as I craned my neck for a last look while continuing to descend alone into the blackness. I directed the beam straight down, but it was not powerful enough to reach the bottom, creating the illusion of an endless, dreamlike fall. Pumping two blasts of air into my buoyancy compensator was not quite enough to slow my downward plunge, and I began a methodical flutter kick of my fins. Shining the light, the only light in my universe, to my console, I watched the depth gauge's black-hand creep by 210, 215 feet, before punching the inflation button on my dry suit, causing the needle to steady up at 220 feet. I breathed deeply.

To reiterate, oxygen is essential for survival, but nitrogen is an inert gas and inactive in the body. When we breathe air, the nitrogen enters body system through the lungs and bloodstream with each inhale, and exits with every exhale. The net sum is a wash. Once air is compressed in a diving tank, and then breathed under the pressure of water above, the physiological effects of both oxygen and nitrogen change drastically.

The underlying reason for the difference between breathing compressed gases underwater and breathing the same mixture on the surface revolve around a concept called "partial pressures." The atmospheric pressure that we live with at sea level is the base measurement that we use for comparing the increasing pressures of depth, and the decreasing, thinning air of higher altitude. Most of us live at sea level in one atmosphere of pressure, which is equivalent to 14.7 pounds per square inch of pressure. When you fill your car or bicycle tires to a certain "psi," or pounds per square inch, you are actually increasing the pressure inside the rubber tube to the desired amount above the ambient 14.7 psi that already exists at the surface of the earth.

Because oxygen comprises 21% of air, its partial pressure in air is 21% of 14.7 psi, or 3.09 pounds per square inch. An English graybeard physicist named Dalton proved this in the early 19th century. "Dalton's Law" states that the total pressure of a gas mixture equals the sum of the partial pressures that make up that mixture.

Pure, 100 % oxygen is highly toxic when breathed at something greater than 1.6 atmospheres of pressure, depending on individual physiology. The surface pressure of 14.7 pounds per square inch multiplied by 1.6 atmospheres equals 23.5 psi. At something greater than 23.5 pounds per square inch of pressure, oxygen will adversely affect certain people's central nervous systems, resulting ultimately in uncontrollable convulsions, and, if underwater, drowning. Pure oxygen reaches 1.6 atmospheres at a depth of about twenty feet of seawater. That is why air is breathed underwater and not pure oxygen. The effect of oxygen toxicity on the human body varies widely among individuals, but there exists an absolute limit to the pressure at which oxygen can be breathed before it becomes deadly. Breathing pure oxygen underwater would limit the average diver to a depth of about twenty feet.

Breathing air does not eliminate the dangers of oxygen toxicity at great depths. Every thirty-three feet of salt water adds one atmosphere of pressure. The deeper a diver descends, the higher the partial pressure of oxygen becomes in the air that he or she is breathing. The partial pressure of oxygen in air reaches the same potentially deadly point as pure oxygen at twenty feet when a diver descends to approximately 220 feet deep. This is not a hard and fast rule, and dives on air have been made to depths far greater than 220 feet, in fact, dives on air have been made in excess of 400 feet. The critical point to remember is that there is an absolute depth limit physiologically that the human body can endure while breathing air. The boundary is ultimately a function of the partial pressure of the oxygen in air, and while the limit exists, it is not written in stone, but can be widely variable dependent upon individual physiology.

Descending 220 feet deep into the Foyer Deck was approaching this limit. To dive near the maximum depth imposed by the risk of oxygen toxicity is to stake one's life on the gambit that an individual's tolerance is higher than the average. Most divers breathing air recognize that guessing a personal depth limit is a crapshoot at best. Most people who dive deep on air respect 220 feet as a physiological limit, and only fin their way beyond it into great uncertainty and risk. Two hundred twenty feet below the surface of the water should not be viewed lightly - this is about twice the depth the average, reasonably accomplished sport diver would consider descending.

At 220 feet deep, in the pitch darkness within the *Andrea Doria's* cavernous Foyer Deck, the narcotic pinging in the back of my consciousness increased noticeably. Each exhale sounded different, I could almost discern the crisp pop of individual bubbles as they left my regulator and accelerated upward. Two hundred and twenty feet is only about ten floors high in a building; it is shorter than the length of a football field and can be sprinted in less than ten seconds. Two hundred twenty feet underwater is a lifetime from the surface. I concentrated on maintaining mental focus.

Partial pressures affect nitrogen as well, causing nitrogen narcosis to grow more intense the deeper one dives. The high partial pressure of nitrogen has an anesthetic affect; it has to do with the solubility of nitrogen in the lipids in the brain. The nitrogen seeps into the fatty structures around the brain and slows cell communication, but no one is absolutely certain why. Many other inert gases exhibit similar characteristics, with helium being a notable exception. Narcosis can manifest itself in divers very differently depending on individual physiology and physical condition. Being cold, tired, or out of shape can significantly change narcosis's influence during otherwise identical dives to the same depth. The rule of thumb is that being "narced" will accentuate the state of mind the diver has at the time. If the diver is relaxed, then being narced may relax them further, in the extreme to unconsciousness. If the diver is anxious, being narced may push them to paranoia and panic. What narcosis will certainly do is to cloud judgment and make sound and timely decisions more difficult.

One can learn to cope with some of the effects of narcosis through experience by building habit patterns and a feel for what is "normal" in an utterly un-normal world. But practice and experience do not physiologically overcome the influence of nitrogen narcosis on the body. Scuba diving classes used to teach an oversimplified illustration of nitrogen narcosis called "Martini's Law." For every fifty feet of depth, it was similar to drinking a martini. Anyone who drinks knows that the alcohol's effect can vary dramatically depending on the individual and their physical condition at the time. Fatigue or hunger can exacerbate the effects of alcohol. Nitrogen narcosis, like intoxication from alcohol, has many influencing variables that make the seriousness of the condition difficult to predict.

"NORMAL" IN AN UTTERLY UN-NORMAL WORLD

The impact of narcosis can quickly become catastrophic because the condition gets worse with depth, which is exactly where divers need to have their wits about them most. It is theoretically easy to alleviate the effects of nitrogen narcosis; the diver simply ascends, possibly only a few feet. Much like being drunk, however, being narced is not necessarily easy to determine by the individual. The number of people that continue to drive drunk is a testament to human judgment in regard to their own sobriety. Narcosis is similar in that there is no consistent, quantitative number to gauge actual reasoning ability. A person might react differently to a common depth under varying circumstances of visibility, warmth, and general comfort. Swimming inside a blackened shipwreck in a silt strewn world of rusty-steel shambles is an intrinsically disorienting experience, and nitrogen narcosis makes it far more confusing.

I panned my light in a broad semi-circle and tried to paint a mental picture of the surroundings. My mind wandered. In a vague, passing riddle I wondered where the bubbles coming out of my regulator would eventually "settle" in their condition of apparent reversed gravity. Would they stream straight up out of the shaft thirty feet above, racing with increasing speed until they broke the surface at seven times their size when exhaled? Or would the light current inside the wreck push them sideways in their initially slow ascent, trapping them in a passageway or space with no vertical escape to the sun? How much 1980s vintage air was trapped in the hull of the *Andrea Doria*, the only commodity divers willingly left behind on their quest of exploration? For that matter, did any 1956 air still remain in some isolated corner of the ship's hull? What would that bare spot free of water look like? It was starting to get weird, a warning sign to break off the line of thought and refocus. I directed my attention to the task at hand.

My goal was to find the *Andrea Doria's* gift shop. Gary Gentile had made an earlier dive into the apex of the collision between the *Andrea Doria* and the *Stockholm* and returned with a treasure trove of costume jewelry. A Vietnam combat veteran and serious adventurer, Gary Gentile was probably the most accomplished east coast wreck diver in the early 1980s. He had described the location to me with admirable candor in his willingness to share information on a "hot spot" for artifacts.

I crossed my fins, pushed against the water, and slowly spun my body around to make certain to face opposite from the direction of the Dish Hole, now out of sight above me. I swam forward and looked for an entrance toward the *Andrea Doria's* bow. It was dark, and the light made a pitiful impact in penetrating the surrounding confusion of disintegrating bulkheads and rubble. I continued to kick forward.

Hovering at 220 feet deep, I tried to recall Gary Gentile's precise words. Damn it. With the same attention span demonstrably lacking in college, the directions were not firmly entrenched in my mind. I thought the surroundings would make the location more obvious once in the general vicinity of the gift shop. The narcosis and lonely darkness didn't help my confusion.

After spending ten minutes cautiously searching the periphery of a maze of fallen debris and cables, I became hesitant to proceed too far from the relative safety of the shaft. Maybe it was the nitrogen narcosis, perhaps common sense, but the nagging unease kept me from pressing in very far. As it turned out, I had misunderstood Gary Gentile's directions and had swum toward the bow when I should have been going toward the stern. *Andrea Doria* blueprints were not common in 1984, and my mental image of where to go was derived strictly from boat-deck conversation. I slowly backed away from the Foyer Deck bulkhead, hands empty for my effort, and regained my bearings before starting up to regain contact with Gary Gilligan at our briefed rendezvous time. The narcosis seemed to ease at 205 feet and I waited at the entrance to the Dish Hole feeling comparatively clear-headed.

A dim light slowly became visible down the silty mess of the Dish Hole, but it was impossible to determine how far away. Gary's form became recognizable, and I could see that he was coming out. I exhaled a slow stream of bubbles, relieved in the knowledge that I would not be forced to look for Gary so late in our dive. My timer clicked over to eighteen minutes on the bottom. It was time to find the anchor line.

"NORMAL" IN AN UTTERLY UN-NORMAL WORLD

GARY GILLIGAN (LEFT) AND THE AUTHOR (RIGHT) DISCUSS THE LOCATION OF THE GIFT SHOP WITH GARY GENTILE (SITTING). FROM THE AUTHOR'S COLLECTION; PHOTOGRAPHER UNKNOWN (1984).

We ascended, exited the Foyer Deck through Gimbel's Hole, and finned the short distance to the anchor line. The contrast of the sun, amazingly bright at 170 feet, was striking. I grasped the rope to the *Wahoo* with my right hand and swung my console up to my face: twenty-one minutes. Damn, just slightly over what I had planned to spend. The tanks on my back held eighty cubic feet each for a total of 160 cubic feet of air. After twenty minutes with a maximum depth of 220 feet, there was not a big margin for error with my air supply during my decompression. I needed to start my hang. Without realizing it, I began kicking faster to the first decompression stop in a counterproductive attempt to expedite the venting of nitrogen from my body. Rising through 120 feet, I felt a wave of heat pass through my temples, followed by a slight headache; I was going up

White arrow points to the Foyer Deck entrance to "Gimbel's Hole." Photo courtesy of The Mariners' Museum, Newport News, VA.

Black arrows depict route to the Dish Hole and First Class Dining Room (top arrows), and to the Gift Shop (bottom arrow). Photo courtesy of The Mariners' Museum, Newport News, VA.

"NORMAL" IN AN UTTERLY UN-NORMAL WORLD

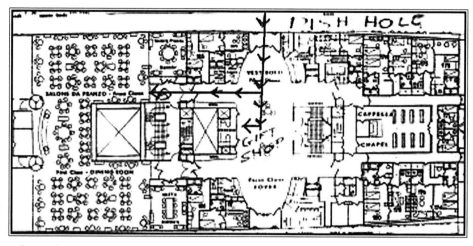

Gimbel's Hole depicted by vertical arrows at the hull. Dish Hole depicted by top horizontal set of arrows going aft (the last arrow stops at the First Class Dining Room). Gift Shop depicted by bottom horizontal arrow going aft. Courtesy of Steve Bielenda.

too fast. I immediately grabbed the anchor line and stopped my ascent. I had gotten caught up in racing the bottom timer's sweep hand in an effort to leave depth before racking up more decompression time, an extremely stupid thing to do. Going up slowly and allowing the nitrogen to vent gradually during ascent and not just at the required decompression stops was critical. The Navy Dive Tables were uncertain enough as it was.

We didn't have computers to assist us with our required decompression profiles in the early 1980s; empirically derived decompression data did not exist. Instead, we relied on the "Navy Tables" to keep us from getting bent, which were more akin to anecdotal procedures for safe decompression than anything derived through scientific method. The tabulated format of depths, time spent at each depth, and required decompression stops for each combination of bottom time and depth were gathered by the U.S. Navy in a bizarrely practical manner. Divers were sent down for a set time to a specific depth, brought to the surface, and if they displayed symptoms of the bends they would be put in a recompression chamber for treatment.

SETTING THE HOOK

The data were correlated with hundreds of other test dives, and the Navy Dive Tables were born. Not an exact science to say the least.

The trial and error testing had a razor thin safety factor. If a diver using the Navy Tables was fat and out of shape compared to the Navy test subjects, then the Tables were less accurate. If a dive was more arduous than it had been for the test divers, or an individual experienced greater cold, fatigue, dehydration, or any of a number of other factors, then avoiding decompression sickness was no longer assured even if strict adherence was paid to the Navy Tables. On top of all this, the Navy Tables did not account for what were commonly called "silent bubbles," nitrogen gas that might form in a dangerous manner but had no symptoms more evident than fatigue or nausea, both common conditions on a boat at sea. The cumulative effect of slight cases of the bends on human bones, tissue, and possibly neurological function was destructive.

I never really appreciated at the time the true hazard of this potential danger. A headache on ascent was an instinctually bad symptom to me, however. I stopped at 120 feet until the pain was gone, and then continued up behind Gary Gilligan more slowly.

My decompression profile was different from Gary's; he had spent twenty minutes at a maximum of 205 feet deep, I had gone down to 220 feet. Gary left each stop slightly before me. Hanging on the anchor line, arm stretched out horizontally in the current, I changed hands often, afraid that a tired arm might slip off the rope and force a swim against the current to regain a hold. I did not want to surface at the end of my decompression hang miles away and probably out of sight from the *Wahoo* crew topside. I spent seven minutes at thirty feet, fourteen minutes at twenty, and then padded my ten foot stop with a couple minutes of Kentucky wind-age and hung for twenty-seven minutes.

Halfway through our hang time at ten feet, a pair of divers brushed roughly by us on their descent, not relinquishing their grasp on the anchor line for a moment in the current. One was clearly Mike Moore, the only diver on the expedition to use a wet suit. Mike possessed a bit of extra thermal insulation naturally, and apparently the cold didn't bother his

"NORMAL" IN AN UTTERLY UN-NORMAL WORLD

stout torso. Mike's buddy was a new guy to the *Wahoo*. The pair continued to descend below us and the new guy diver's red coveralls slowly faded from view.

I shivered in my leaky, used dry suit; it was the only one my college budget could afford. Toward the end of my required decompression hang I switched to my small, backup pony bottle to avoid sucking the main tanks dry and possibly cause corrosive salt water to be introduced into the cylinders and regulators. Twenty minutes at 220 feet deep with a total decompression of forty-eight minutes. It seemed like an eternity in the cold, wetness of my "dry suit."

I let go of the anchor line, allowed the current to push me under the *Wahoo's* keel, and carefully guided my drift with modulated kicks to stay directly under the boat until reaching the welded steel ladder hanging five feet underwater from the dive platform at the *Wahoo's* stern. Keeping my fins on, I awkwardly climbed up the widely spaced rungs onto the swim platform. If a diver took their fins off before getting out of the water, they ran the risk of being unable to swim against the current if they inadvertently fell back in. I crawled onto the deck on my knees, reached back and removed my fins.

Exhausted, but anxious to get out of my soaking wet long johns and into some dry clothes, I welcomed Hook's help in removing my twin tanks. He carefully put the heavy package down on the deck so as not to crush the regulators or instrument console. It was important to keep up momentum and steadily remove and stow gear before another diver surfaced, only to find their way onboard the *Wahoo* blocked. I used a sturdy bungee cord to secure my tanks to the rail of the *Wahoo* next to dozens of others, some full, some already used. Methodically I put my assorted loose gear - mask, fins, mesh bag, lights, knives, reel, and lift bag - into an oversized nylon bag. I took off my weight belt with the attached five-pound hammer brought along for delicate work, and placed them out of the way in the milk crate behind the *Wahoo's* topside ladder.

Steve looked down at me from the bridge just as I turned my "dry suit" inside out and emptied the gallon of water from the low points of the legs.

It would be close getting the sun to dry it out before this afternoon's dive. He shook his head and the creases in his full, hardened face broke into a broad smile. "Cheap college kid," he must have thought. I turned away, still shivering in the warm sun, and began to get out of my sopping wet long johns.

With unmistakable urgency a shout came from up forward; it was Steve's voice, "Diver up off the bow!"

CHAPTER TEN

"We're having some fun now."

July 15, 1984, fifty miles south of Nantucket Island, Massachusetts

Off the bow; the current was pushing into the bow and the wind was in line with the current. How the hell had this guy come up off the bow? If the diver had lost the anchor line during ascent, he should have come up well aft of the boat after decompressing. The distant figure was difficult to see clearly, laying low in the gently rolling swell seventy-five yards away. The diver was not moving.

Craig and Hook practically jumped over me, hauled in the inflatable chase boat tied to the stern of the *Wahoo*, and pull started the outboard. In less than thirty seconds they were racing flat out to the diver floating in the ground swell off the bow. The cabin emptied onto the deck as passengers and crew crowded the starboard rail, straining to catch a glimpse of events in the distant patch of water.

Captain Janet stood on the topside walkway and pointed her arm toward the floating diver in a constant directional fix to help Craig and Hook, both unable to see the floating diver from their sea level perspective. I even lost sight at times from my vantage six feet higher on the deck of the *Wahoo*, but Janet was standing on the bridge more than twelve feet off the water and could maintain visual contact with the still figure. Craig faced backward into the bright morning sun, shielding his eyes so that he could see Janet's outstretched arm and give Hook small heading changes until

SETTING THE HOOK

they were close enough to keep sight of the diver. It was a beautiful day; this kind of shit wasn't supposed to happen on beautiful days.

The chase boat came to a sudden halt and blocked our view. For what seemed the longest time, but couldn't have been more than a minute, all that was visible were Craig and Hook's backs as they leaned over the inflatable's right pontoon. Seconds later, they hauled in the limp form and dropped him unceremoniously down onto the boat's floorboards. The diver was wearing red coveralls. Craig's head disappeared from view below the inflatable's pontoon while Hook pulled in the diver's floating tanks; for the time being they were the only clue to what had gone wrong during the dive. The chase boat kicked into gear and shot back to the *Wahoo* with the outboard engine screaming wide open. I waited impatiently with growing foreboding for the few seconds until the inflatable reached the stern.

Five passengers and crew were on the dive platform at the *Wahoo's* stern when Craig and Hook arrived with their grim package. Hands eager to help reached out and pulled the chase boat in snug, holding it tight against the pull of the current. Craig and Hook stayed in the inflatable and awkwardly balanced against the swell while they lifted, pulled, and tugged at the lifeless form, trying to move him to the aft deck of the *Wahoo* where it would be easier to work on him. Craig shifted his grip around the diver's shoulders, pinched off his nose, and planted his mouth over the diver's limp lips in a rescue breath. Craig raised his head and began to help with the lifting again. There was blood in his beard. This was not good.

Steve quickly directed the effort on the stern, and seconds later the diver was being hoisted over the *Wahoo's* aft railing. It was Frank Kennedy, a first timer on both the *Andrea Doria* and the *Wahoo*. An instructor, he had come down from Massachusetts for the ultimate dive. I helped lower him quickly on the rigid fiberglass deck and realized that the mouth-to-mouth resuscitation needed to continue. I had never done this before, but Steve required that all crew on his boat were CPR certified. I lowered my head and began to give mouth to mouth just as the Red Cross had taught us. This was not going to be fun.

"WE'RE HAVING SOME FUN NOW."

It took me a few tries before being able to consistently make a leak free seal - Frank's face was completely limp and his relaxed cheeks gave way to my attempts. Rick Jaszyn verified the absence of a pulse and began chest compressions. Other than Rick's figure above and to the right of me, I was in my own world, listening to Rick count out the compressions, waiting for five before breathing more air into the diver's lifeless body. The instant my exhale into Frank's mouth ended, I would yank my face away, but never quickly enough. Each pause in a rescue breath allowed a blast of stale air to shoot back into my mouth from Frank's damaged lungs, a blast of twice used air bringing with it blood, vomit, and salt water froth. I tried to zone myself out to keep from gagging, tried to become mechanical and not think about what was happening.

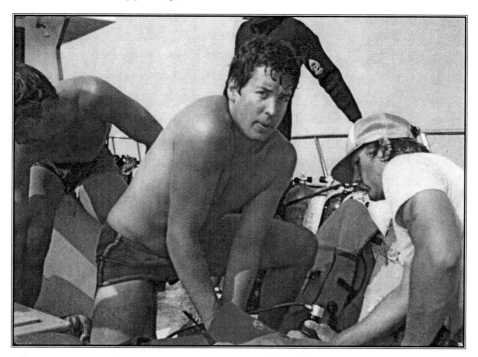

The author gives chest compressions while John Moyer administers oxygen. Gary Gilligan is in the background. From the author's collection; photo taken by Steve Bielenda (1984).

"One, two, three, four, five," I gave another breath, pulled away, and felt blood dripping slowly down my cheek. The counting started again and I concentrated on the routine, barely sensing the buzz of surrounding activity. Rick Jaszyn took a break, and John Moyer resumed with the heart compressions.

On the *Wahoo's* deck the passengers were temporarily consigned to frustrating inactivity, watching events unfold, helpless to influence the outcome. Steve yelled out to Janet on the bridge to get the Coast Guard on the radio, to initiate the lengthy process of a helicopter rescue a hundred miles out at sea. Steve then ducked into the *Wahoo's* cabin and returned to the aft deck with a large green suitcase containing an oxygen cylinder. He opened the case and finished setting up the oxygen ventilator.

I concentrated on the task at hand.

"One, two, three, four, five," I lowered my face, pinched off Frank's nose, and turned to breathe life into his loose form. I raised my head, made a futile attempt to wipe the traces of blood covering my chin onto my bare shoulder, and then saw with relief a bunched up towel in front of my face. Gary Gentile leaned over my back, gave my face a quick wipe, and then made a cleaning sweep of Frank's face. I waited for John Moyer's heart compression count to reach five and kept my hands in place by the Frank's head. Gary Gentile remained fixed at my shoulder, cleaning two faces sharing one lungful of breath. What a hell of a difference that made.

A few, long minutes later I saw a clear mask at the end of a green cord drape its way lengthwise up the diver's body and over his shoulder. I made one last rescue breath, and then placed the oxygen mask firmly over Frank's face and waited for John Moyer to count to five. Gary Gentile gave my face a final wipe with the towel, and we resumed the ventilations with 100% oxygen, the best treatment response we could provide an injured diver out in the ocean and far away from a recompression chamber. Pure oxygen would better feed Frank's starved body traumatized by what we could not determine for certain was an air embolism, the bends, drowning, or all three. One hundred percent oxygen, over time, would also create a pressure

"WE'RE HAVING SOME FUN NOW."

differential across the body tissues and help to shrink the nitrogen bubbles coming out of solution in Frank's body.

Steve and Craig lifted Frank's calves up onto a nearby cooler, attempting to keep his brain and heart at low points of travel for any bubbles that might be emerging in his system. Craig turned to the cabin for a moment, and then reappeared with a pair of heavy-duty scissors. The chest compressions stopped momentarily as Craig cut off Frank's neoprene dry suit, starting with his tight rubber neck seal, and then cut lengthwise at the chest to remove the elastic barrier between the heel of John Moyer's hand and the heart he was trying to coax back to movement.

Craig looked up at me, my hands cupping the green oxygen regulator at Frank Kennedy's relaxed mouth, and said without expression, "We're having some fun now."

We were bound by unspoken duty to repeat that phrase whenever things particularly sucked. This qualified. "Some fun now." I replied.

Steve reached down, grasped Frank's right hand, and searched for a pulse. He shook his head from side to side, and we continued with our grim routine.

Minutes ticked by, then an hour. Before the Coast Guard helicopter arrived the oxygen ran out, and for the last fifteen minutes that the hapless diver lay on the *Wahoo's* aft deck we took turns with the mouth-to-mouth resuscitations.

Finally, a speck on the horizon transformed into the orange and white SH-3 Coast Guard helicopter. None of us had any illusions as to Frank's chances for survival by now. Early on in the effort Gary Gentile thought he had felt a very weak, fleeting pulse, but that was over an hour ago, and the faint sign had vanished quickly and permanently. We were under a moral and legal obligation, however, to continue with the resuscitation efforts until relieved by the appropriate medical authority. We all breathed a sigh of relief as the helicopter approached the stern and instructed Janet in the wheelhouse by radio to steer the *Wahoo* slightly port of the prevailing wind. The helicopter slowly entered a hover over the *Wahoo's* clear aft deck. Earlier, Steve had directed that all loose dive gear be stowed in the cabin or up forward, away from the rotor wash. A long, orange rope touched the *Wahoo's* deck, and John Moyer, one of the five passengers and crew Steve had selected to assist with the delicate operation, grabbed it and began to haul in the metal litter basket with its orange flotation cells. He made certain to ground the attached steel cable to the *Wahoo's* railing before placing the litter flat on the deck in order to disperse the built up static electricity from the helicopter's rotors.

Craig finished the final duties before loading Frank in the litter. He used the scissors to cut off the dry suit from his arm, and in thick, black magic marker wrote down what little information we knew about his dive profile: "Depth 240, 25 minutes, ran out of air." Frank had sucked his main tanks dry and the maximum indicator needle on his depth gauge read 240 feet. His bottom timer showed twenty-five minutes. Theoretically the depth and circumstances would assist the medical personnel at the recompression chamber in their attempt to revive the unresponsive diver. We all hoped, but I doubt any believed, Frank would be put in a chamber and resuscitated. It was just too late.

The helicopter raced with futility to the hospital in Rhode Island. Frank Kennedy was pronounced dead soon after landing. One and a-half-hours had elapsed since he surfaced, which I considered pretty quick for a helicopter crew to gear up and sortie out this distance, but still far too late. Such was the additional risk of diving the *Andrea Doria*, over a hundred miles from the nearest medical facilities with a recompression chamber.

After the Coast Guard helicopter pulled in the rescue litter and sped out of sight, information began to trickle in as to what had happened underwater. It would never be enough to determine the exact circumstances, only Frank would know that for certain, but there existed sufficient information to make some educated guesses.

Frank had come out on the *Doria* trip solo, planning to hook up with a buddy from among the passengers on the *Wahoo*. His dive buddy was Mike Moore, one of the regular passengers on the dive charter. They had conducted a semblance of a sit-down, face-to-face brief, discussed their underwater goals, and had dived together the day before. Still, they were no more than casual acquaintances, and the first dive experience they had shared was on the *Andrea Doria*. Deep diving procedures were by no means standardized in the early 1980s; Frank was certainly taking on a lot of unknowns on this second dive to the *Andrea Doria* and his first time with the intent of penetrating the wreck.

But any discomfort that Frank Kennedy did, or did not, experience was certainly not Mike Moore's fault. Mike was in a similar position, with no predetermined partner on the *Wahoo*. The only difference was that he was more experienced diving wrecks and was used to how the *Wahoo* operated. Mike was committed to a self-sufficient diving style and ethos, and a buddy merely added to what he already considered was a reasonably safe diving profile. It would become evident that Frank had not achieved the same degree of self-sufficiency. This, in part, is what killed him.

In contrast, our first several dives on the *Doria*, while not overly cautious, took the fullest advantage of the buddy system and self-sufficiency. The first four dives that Hook, Craig, Gary, and I made to the *Andrea Doria* the previous year were as a four-person team. We each may have strayed away from the nucleus at various times, but the additional divers and air we carried translated into options if something went wrong. None of us had dived deeper than 160 feet before our first *Andrea Doria* dive, there was no formal training available, and it was learn as you go. But we each had much experience diving other wrecks in the same waters, and we each knew exactly what to expect from our dive buddies.

On our first two dives on the *Andrea Doria* we went into the Dish Hole, but only a pair of us entered at one time. The other two waited at the Dish Hole ledge in the Foyer Deck shaft, shining powerful dive lights inward as a reference for the scavenging divers. Although the lights were not visible inside the shaft through the thick cloud of silt, we knew as the digging began that it would only be a short swim before we would see the reassurance of our buddies signaling that help was available if needed. We broke up into smaller groups on subsequent dives, but our familiarity with each other during those critical first few *Doria* dives helped immeasurably.

SETTING THE HOOK

During our first expedition, Gary had lost the anchor line and made too rapid a descent swimming hard into the swift current. Once on the wreck and now completely disoriented by the exaggerated effect of the narcosis, Gary began to swim in a constant, tight circle to the left barely above the *Doria's* hull while the three of us watched with amused concern. I finally grabbed his arm and got eye contact. Gary may not have known where he was or why he was there, but he knew to trust us. We guided him back to the anchor line and sent him up for decompression after only a few minutes on the wreck. What could have been a disaster became a learning experience. It was also a humorous anvil we could hold over Gary's head the next time we shared a Budweiser.

Frank Kennedy could never have achieved a similar level of intuitive understanding and trust with a dive buddy he had only just met. Neither diver would be capable of detecting the subtle warning signs of debilitating narcosis in their buddy, since neither was familiar with how the other diver acted under normal conditions. Mike Moore and Frank Kennedy had planned on going no deeper than 205 feet in an exploration of the Dish Hole. Soon after passing Gary and me in the opposite direction during our ascent, at just seventy-five feet deep, Frank had raced toward the wreck causing Mike Moore to lose sight of him almost immediately. He would not see him alive again. Frank probably entered the total blackness of the Foyer Deck shaft and was possibly looking for a treasure trove of china another diver had lost on an earlier dive.

Big Artie Kirchner had accidentally dropped a mesh bag jammed full with china down the bottom of a stairwell halfway along the passageway leading to the First Class Dining Room. Artie - an experienced diver respectful of his limitations - did not pursue the artifacts into the unknown darkness and depth after losing the bag. Back onboard the *Wahoo*, the other divers quickly learned about the dropped bag and perhaps Frank simply could not resist trying to make what seemed to be an artifact coup. Never having been inside the *Andrea Doria*, it might have seemed like an easy task to Frank as he stood on the deck of the *Wahoo*, but the tone of Artie's voice made it clear to me that it had been a hairy enough dive as it was,

"WE'RE HAVING SOME FUN NOW."

and venturing through the narrow opening into unknown and unplanned depth would have been plain stupid.

Mike Moore looked for Frank only as far as they had planned to go into the hulk of confusing wreckage, and then waited at the top of the shaft to the Foyer Deck as long as his decompression schedule would allow. When a plan went this far awry, it was a diver's responsibility to protect his or her own life first. With mounting concern, but no good options, Mike ascended for the long decompression solo. By the time he surfaced his worst case scenario was realized and he climbed onto the *Wahoo's* deck after the drama had been unfolding for more than forty minutes. The remainder of the story could only be pieced together.

Frank Kennedy more than likely became lost inside the *Andrea Doria*. Where exactly he had gone would remain a mystery forever. That he eventually found his way out, perilously low on air, after descending to 240 feet within the wreck was amazing. He obviously did so at the absolute last opportunity - he did not even have enough air remaining to make the short swim to the anchor line.

At some point, perhaps once out in the light on the hull of the *Andrea Doria*, probably at 170 feet deep, Frank Kennedy ran out of air. The hard to draw pull of the last few breaths from his regulator must have alerted him, must have scared the shit out of him with the panicky feeling in his chest of impending suffocation, worsening as each determined kick of his fins toward salvation depleted the oxygen in his body. Frank Kennedy faced a surreal choice between two impossible options; he was trapped between the deep-seated understanding that a rush to the surface would be disastrous, yet had to be even more certain that remaining underwater would mean death by drowning in a matter of seconds.

The tragic irony of the event was that a post accident inspection of his dive gear revealed that his emergency, backup pony bottle was completely full. The regulator attached to it, however, was routed around his main tanks in a way that allowed the regulator to fall out of his reach. Steve had recommended to Frank that he re-route his pony bottle after his first *Doria* dive, advice obviously not heeded. The life saving air that was literally inches behind Frank's head was inaccessible, forcing him instead to make an out of air ascent to the surface.

The doctor would later determine that Frank died of an air embolism. The rapidly expanding air could not vent from his lungs quickly enough. Instead of exiting Frank's over-inflated lungs through normal exhalation, the gas bubbles passed back through the lining of the lungs into the blood stream. These bubbles continued to expand in the lessening pressure in Frank's rush to the surface, probably gathering in the arterial system and traveling to the brain. Thankfully, it was all over quickly; Frank Kennedy was likely dead when he hit the surface.

In the ad hoc university of on-the-job experience that was our fundamental guide back then on how best to dive a deep wreck, this event was a full course. There was precious little new to learn from the tragedy, but the lessons revisited were reinforced in concrete. Know your dive buddy, stick to your dive plan, respect your limitations, and perhaps most importantly, be 100% positive that all redundant backup systems can be accessed effortlessly. Frank's full pony bottle sent chills down my spine.

As an immediate result of the circumstances surrounding Frank's death, I decided to donate a spare regulator, and Steve offered a tank, to place on the wreck at the anchor line's shackle for emergency use by an ascending diver low on air. In the unpredictable weather and current anchored over the *Andrea Doria* it was possible that the *Wahoo* might break free at any time. If this happened the tank and regulator would be lost forever. It was a small risk in the value of equipment though, even to a broke college kid, if a diver needed the air. Occasionally on wreck dives a spare cylinder for use as an extra supply of air during decompression would be tied off to the anchor line at ten feet deep. We had not considered the more pressing possibility of requiring the extra air while still on the shipwreck's hull. Now we knew.

After the Coast Guard helicopter departed with the Frank Kennedy's body, we all sat quietly in the eerie calm trying to decide what to do next. The hastily stowed dive gear was gradually retrieved from the *Wahoo's* cabin and reclaimed by the various owners. Steve left it in the hands of the

"WE'RE HAVING SOME FUN NOW."

passengers on how next to proceed. A majority show of hands by the group opted to stay out, to make the second dive planned for the day before returning to port at Montauk, New York.

My initial inclination was to skip the dive altogether, but a dive trip to the *Andrea Doria* was not easy to pull off, and we had all risked a lot in time and money simply for the opportunity to dive the wreck. An ominous pall hung over the *Wahoo*. This was not the type of event that quickly receded to memory. The passengers decided to compromise, to finish off the charter with one more dive, but to cut the trip short a day and head home that evening.

Seven hours later, Craig and I were back in our dry suits, swimming down the anchor line. Craig retrieved a Promenade Deck window that dive, but I have no memories of the event; my only evidence of his find was a notation in my logbook. My only lasting memories of those two days in 1984 over the *Andrea Doria* are of my attempt to find the ship's gift shop and Frank Kennedy's death. The ten-hour run back to Montauk was unusually quiet. Most drank beer, most a lot, but there was a marked lack of spontaneity and light heartedness to the usually raucous return home. Not a big surprise.

I did learn a little more about Steve Bielenda, however. Hook, Craig, Gary, and I were standing at the bow of the *Wahoo* the following week on the season's second expedition to the *Doria*, planning our next dive and taking in the fresh, salt breeze in the relative solitude forward. Steve walked up, and in an uncharacteristically quiet, almost shaky, voice said:

"Hey, you guys be careful; really."

The words were simple, but the tone was clear. He turned his back, but not quickly enough to prevent us from seeing the glint of moisture in his eyes. This was a side of Steve that I knew existed, but had never seen. It got our attention. For all his brusqueness in the rough and tumble world of the east coast dive charter business, Steve was human. He definitely gave a damn.

Over the years this was how we learned best to conduct our dives; hopefully through word of mouth, too often through personal experience and tragedy. There were no courses to teach what we were doing, no books to use as reference for our exact intent. Each lesson provided a painful data point to further our collective knowledge and build a haphazard learning curve of individual experiences ingrained deeply in the psyche. When I left northeast wreck diving in 1985 these were our only guides.

I was soon to learn how far things had come.

FROM LEFT TO RIGHT: CHRIS DILLON, HANK KEATTS, GARY GILLIGAN, CRAIG STEINMETZ, DON SCHNELL (HOOK). TIM NARGI SITTING. OUTSIDE THE *WAHOO'S* WHEELHOUSE ON BLOCK ISLAND. FROM THE AUTHOR'S COLLECTION (1983).

CHAPTER ELEVEN

Mistake number three...

October 2000, Whidbey Island, Washington State

The gradually sloping silt-covered bottom, barren except for a peppering of fist-sized rocks, dropped away almost imperceptibly. Pausing in my easy kick down, I turned to face Ron Martin, raised the camera with its small target light fading into the receding darkness, and snapped a shot. The flash was blinding in the surrounding blackness, but our eyes readapted quickly with the upward swing of dive lights. I pointed the cylindrical light at my wrist computer; 162 feet. I felt positively euphoric, not narced, but completely at ease and comfortable in my weightless hover surveying the silty bottom. It had taken until my fourth deep practice dive, but my movements now seemed effortless. With a deft twist of fins I swung around 360 degrees until facing Ron once again.

Ron gave an inquisitive thumbs-up. I responded in kind with certainty in silent agreement that it was time to start up. In unison, we turned toward the slowly rising bottom that led to shore and felt the unexpected light push of a current against our chests. There was supposed to be another forty minutes of slack tide at our chosen dive spot on the north side of Allen Island, but judging the exact tidal swing in each cove, nook, and cranny in the dozens of area islands and rock outcroppings was not an exact science.

We had anchored the SeaRay in fifteen feet of water and reasoned that we could dive together, that there was no need to leave anyone in the boat

SETTING THE HOOK

while it was securely anchored in the shallows. After a minute of kicking vigorously into the growing current and only ascending to 155 feet, it became apparent that we had thought wrong. For a reason that I didn't waste time trying to fathom, the down-welling current had decided to pick up at our deepest point on the dive and was now trying to push us away from land.

I concentrated on maintaining a steady, even kick, exchanged glances from my console compass to Ron's shadow over my right shoulder, and pulled by hand when a lone rock presented the opportunity. The current got stronger, requiring faster kicks and forcing me to work harder than was safe, but it seemed the only way to make progress up the muddy bottom slope. I looked at the wrist computer's digital depth read-out, saw 140 feet, and simultaneously felt Ron tap me sharply on the right shoulder.

This time, Ron gave a forcefully urgent thumbs-up with his gloved hand; he wanted to go up right where we were, away from the shore and the boat. Damn, I hadn't thought this through. Ron was an experienced diver with sound judgment, but he was overweight and out of shape. He had also given his right knee a good bang on the steel dinghy davit attached to the boat's swim platform while maneuvering to get his tanks on twenty minutes earlier. My swimming pace must be running him ragged. Damn - my first mistake.

I reached for Ron's console and saw that his single, high volume steel tank was over half full. He also had a thirty cubic foot pony bottle strapped to his back. I pointed my hand decisively toward shore, a gesture of, "No, let's press on along the bottom back to the boat." Mistake number two on my part.

We both had plenty of air to make the slow swim to the boat along the bottom against the current, I reasoned, but it would be impossible to beat the same current on the surface and make it back to the boat. Better to stay low along the bottom - my computer indicated an eight-minute decompression hang and the current could move us far, far away from the boat and land if we decided to make a free ascent from our existing position. I gripped Ron's left bicep tightly, made a quick scan of the compass,

MISTAKE NUMBER THREE...

and powered my legs into motion with Ron in tow. Mistake number three - big mistake.

The increased exertion from kicking into the current and attempting to pull Ron was bound to catch up with me. There was plenty of air in my double tanks and we were not that far into mandatory decompression, but I didn't want to lose contact with the boat and possibly end up floating on the surface, trying to hail down a ride back to shore as the current took us out to sea.

After about three minutes of intense kicking and pulling, my eyes squinted to make out the digital read out on the dive computer - 130 feet. Why did I need to squint, I thought? In a sudden rush it all caught up with me, my breathing became labored and it felt as though I could not get enough air. I seemed to be suffocating. Next came the dizziness, with lights sparkling and flickering in my peripheral vision, before everything finally went dark. The sensitive nerves in the eyes were the first to go.

That dreaded panicky urge to shoot to the open air of the surface poked its head at the edge of my consciousness. Oh shit, shit, shit, what the hell have I done? My mind reeled and I desperately fought to stay conscious. I tried to close my eyes, but couldn't tell if I was successful; it was total darkness either way. My overexertion had brought me to the edge of catastrophe, to the point where physical control over my body began to slip. Hold on, don't pass out or you're dead - I repeated the words over and over in my head, let go of Ron's arm, slowed my kicks, and concentrated with all my will to regulate my breathing. Damn it, there's enough air, fight it, fight it, you have enough air in each breath to live; it only feels like you are drowning. I felt wretchedly alone, even my mind was attempting to desert me. Don't leave the bottom, keep conscious, damn it; keep fighting.

The overexertion combined with the resistance of breathing air through a regulator at depth had ratcheted up the carbon dioxide in my system dangerously. The intense activity and physiological effects of breathing air under pressure vastly increased the possibility of a deep-water blackout. I needed to do something; I needed to do something right now.

SETTING THE HOOK

Deep, even breaths, I tried to focus; breathe, just breathe. The words seemed impossible to believe, but still, I kept repeating "There is plenty of air," over and over in my mind, trying to calm, trying not to think of the 130 feet of water between me and the surface. Breathe. It will get better as the carbon dioxide leaves with each controlled exhale. Concentrate, don't relax your jaw, keep the regulator in your mouth, if it drops out you're dead. Stay put, at depth; fight the urge to bolt to the surface.

It probably took no more than a minute, although it felt like forever, before finally beginning to regain full control of my senses. It suddenly got light, my vision cleared and awareness of surroundings widened. A quick check of depth and tank pressures showed 120 feet and plenty of air. Remembering Ron, I turned around; he was gone. Damn it. What had I done? I was fairly certain Ron had decided to surface while still in control, with or without my thumbs-up of approval. But I was not positive, and just had to hope that he had enough control to stop during his free ascent to the surface to decompress.

During a free ascent it was critical that a diver have complete buoyancy control. Without a tangible reference like an anchor line to grab, the only thing stopping an uncontrolled race to the surface and possible decompression sickness or embolism was the gradual and exact metering of air from the diver's buoyancy compensator and dry suit. Each foot a diver went up increased the volume of gas in both as the water pressure dropped and the air within them expanded. It was not difficult to maintain control so long as one did not ascend too quickly and find they were unable to vent air rapidly enough. But it did take practice, and I had to wonder when Ron had last made a free ascent.

I tried to think logically, to reassure myself that Ron had probably already ascended, after all, less than a minute prior he had been plainly visible, but the consequences of the alternative allowed me no peace of mind. For Ron to disappear that quickly on the slowly sloping sea floor presented only two possibilities - either he went straight up and surfaced, or he swam deeper with the push of the current. If he had gone deeper in a disoriented swim he was likely dead by now. The increasing narcosis combined with

MISTAKE NUMBER THREE...

the carbon dioxide buildup of exerted swimming could cause him to pass out and drown; it happened periodically in these waters, so deep and close to shore.

The bottom terrain became visibly steeper as the shoreline neared and the current began to ebb. I slowed my ascent and grew confident that at least I would make it back to the boat. My mind began an anxious wander. Where the hell was Ron? Why did we go diving without leaving someone in the boat? We both knew better, and we rarely left the boat unattended. If someone had been waiting in the boat it would have been a no-brainer to simply make a slow, controlled free ascent, taking our time to decompress hanging weightless just below the surface. Then when we surfaced, all we would have needed to do was wait for the boat to pick us up, not particularly caring where the current had moved us.

If nothing else, I should have let Ron ascend solo when he gave the first thumbs-up. Assuming that I made it back to the boat, there would be someone to look for him on the surface. There had been no current when we jumped in the water, and it was forecast to stay slack for our entire dive. Still, I should have anticipated this, should have headed off the situation long before it became dangerous. For a brief, few seconds I had not been in control of my body, and now I had no idea if Ron was alive or floating face down in a mass of dive gear on the surface or prone on the sea floor. I settled my knees on the rocky bottom at twenty feet and tried to think while the two dive computers counted down the mandatory decompression time. Was there anything that could be done this late in the game?

I ascended to the final stop at twelve feet with eyes glued to the computers, willing the mandatory decompression to end quickly. I thought about going up immediately to search for Ron, but reconsidered. I would be little help if bent, and if Ron had not survived the free ascent - if that was indeed what had happened - then it was too late anyway. Damn. How could I have let this happen?

A shadow on the surface caught my eye and I slowly looked up. The boat had swung on its anchor line and it was nearly directly above me. The yellow of Ron's pony bottle was clearly visible twenty feet away, attached to his buoyancy compensator and tied off to the swim platform. He must be on the boat; he was alive. I decompressed for several minutes longer and concentrated on being patient. There was lots of time later to figure out exactly what had happened, with Ron safe on deck there was no point in pushing limits now.

Finally, I went up and floated at the dive platform while simultaneously shouting questions to Ron. He had made a slow ascent and managed a minimal decompression. The current was running directly to the shore on the surface before hitting the island, at which point it created a down-flow in the opposite direction underwater. After surfacing, the reversed current had taken Ron effortlessly straight to the boat.

This was not the first, or even the second, time I had come close to passing out from "over-breathing my regulator." That was the popular term for the feeling of being unable to get enough air, of helpless suffocation, knowing that a race to the surface would vastly decrease your chances of survival. There was that typewriter on the *Andrea Doria*, where hauling it vertically thirty-five feet with Craig Steinmetz had almost caused me to lose control. There was another time as well; while setting the hook on the *U-853*, I had experienced the same frightening lost control. The anchor had missed the German submarine and fallen seventy feet short, and in the excellent visibility I decided to haul it to the wreck. I pulled the heavy Danforth sand anchor with its thirty foot chain across the sandy bottom and followed the compass course in the probable direction to the wreck. One hundred thirty feet above me Steve was doing his best to keep the *Wahoo* in position and allow slack in the anchor line.

I swam hard, lifting the anchor and humping it a few feet at a time, and was just about to give up, ascend, and tell Steve to try again when the silhouette of the submarine slowly appeared about forty feet away. Only twenty feet from the steel tomb the exertion caught up with me, and I experienced the same horrible paradox as on the dive with Ron; Going straight up would mean breathing even harder in the ascent and increasing the risk of an embolism or the bends. To stay down took tremendous self-control, fighting the overwhelming sensation of suffocation, as I slowly, excruciatingly slowly, caught my breath. Being unable to quickly control my breathing would result in drowning; it was that simple. Eventually, I regained control and cautiously made my way the final twenty feet to the *U-853* to set the hook.

So far luck had been on my side and my brushes with deep-water blackout had not ended in tragedy. The prospect concerned me at least as much as any other risk of diving the *Doria*, like getting snagged by a net or lost inside the wreck. Both had happened to me before, and both were manageable with a relatively clear head (at least as clear as one could expect while narced). The thing that completely scared the shit out of me was the excessive carbon dioxide buildup of "hypercapnia." Of course, there were warning signs, but they were easy to miss. The feeling of losing your vision, your thought process, your very consciousness coupled with the realization that if you did pass out you would most definitely drown was terrifying. There were no good options once a diver approached losing control from excessive carbon dioxide. Prevention was the only acceptable remedy.

We talked a lot about that dive; I thought a lot about that dive. Most of my instincts underwater once in duress were sound, but I was not fully cognizant of my breathing pattern at all times, and my pre-dive planning was awful. I needed to close the gap, and concentrate on the fundamentals underwater. It was unlikely the *Doria* would give me a second chance.

I was not as comfortable with deep dives as I had been at twenty-one. Back then, a natural competitiveness and frequent diving made the depth seem to be not that big a deal. I could afford to be somewhat cavalier back then, but not anymore. My best strategy to counteract my lack of recent deep diving experience would have to be the incorporation of the planning, training, and discipline I had learned flying jets in the Navy. If the process could be broken down to a few, simple principles, then there would always be a semblance of a safety net and a higher awareness to detect when things began to go wrong. I was already doing this intuitively, but now had to raise the principles to the forefront of my consciousness. I came up with three: breathing, buoyancy control, and situational awareness, or "SA." It was nearly impossible to qualify all the hazards of diving, but I thought these three basic concepts of control would head off the vast majority of them.

Above all, I needed to be in absolute control of my breathing, always. If my breathing got away from me at depth it might only be seconds until disaster; the dive with Ron reinforced this basic lesson with frightening effect. Consistently perfect buoyancy control was also critical; if I was working even slightly in an ascent or descent it would cause me to breathe harder and make it more difficult to control my physiology.

Finally, I had to have 100% certainty of my exact situation and what my next plan of action would be underwater - "situational awareness," as it was called in the flying business. I was used to constantly updating my "SA" in the three-dimensional airborne world and now needed to do the same while diving. Take nothing for granted and strive to anticipate the effects of each action before the fact. My goal was to strive to translate the professionalism that possessed my every move in a jet to the world underwater.

Having a somewhat concrete plan helped my confidence, but the dive also renewed a healthy respect for the deep that had begun to slip. I had been scared for myself and for Ron. I wanted to watch my kids grow up. This had to be a lesson I never forgot.

The primary reason to make deep, air dives was to bring back the confidence and proficiency enjoyed years ago. Confidence and proficiency were meaningless if the dive killed me just the same. I made one last, very cautious, deep dive breathing air so as not to end my initial practice sessions on such a disturbing note, this time alone. I would not allow my recreational dive buddies to accompany me on a deep dive again. That had been really stupid.

My solo trip down the cliff face to the blackened void at 207 feet was uneventful, at least as routine as any trip into a sensory depriving abyss of nothingness can be. Hanging neutrally buoyant on the wall, trying to keep rhythm with the steady narcotic pinging of bubbles, my surprisingly high comfort level vanished instantly when I turned off my light. No differentiation between eyes open or shut; no sound but the disturbingly familiar

MISTAKE NUMBER THREE...

gurgle of my regulator and the narcotic pings and pops; all motion slowed to a nightmare-like quality of helplessness under the water pressure and bulk of my gear; I could not have screamed if I wanted to. I endured for a slow ten-count, flicked the light back on, and savored the five foot visibility. Complete darkness was not normal, but I had to make it feel normal. Years ago, it had seemed that the isolation was almost fun. Not anymore - I could do it, but would need to brace for that instant of spine chilling panic the moment the lights went out. My experience lost in the *Andrea Doria's* First Class Dining Room was still fresh in my subconscious. It was time to take advantage of the training and technology made available to sport divers in the last fifteen years.

I started researching how best to fully incorporate the newer, and hopefully safer, methods of deep diving into my old style of operating. I had rediscovered the old way of diving deep, at least as close as was possible separated by all those years. Now I needed to learn the best way.

CHAPTER TWELVE

Breaking the rules

The need to decompress and nitrogen narcosis do not put an absolute floor on the depth a human can endure while breathing air and there is no clearly defined line or limitation. Because decompression and narcosis affect divers differently depending on both individual physiology and circumstance, to venture beyond universally conservative standards - such as recreational depth limits - is to rely almost entirely on experience to mitigate risk. To judge personal limits accurately, one must have experience; to gain experience, one must push the limits. The contradiction is clear.

Actual deep diving was undoubtedly the fastest type of refresher training, but it was also the most dangerous. To simply ignore sport diving's technological advances in the past fifteen years would be plain foolish, and in November of 2000 I drove the hour to "Adventures Down Under," a dive store in Bellingham, Washington, to speak with the owner about the advanced course offered in "technical" and "trimix" diving, both terms I had only heard a handful of times.

I had been standing for five minutes at the store's counter describing my diving background to Ron Akeson, the owner and technical instructor, when a book set on the shelf over his shoulder caught my eye. "The Technical Diving Handbook" was written by Gary Gentile, the near legend of wreck diving of the early 1980s, the guy who had wiped the blood from my face during the failed CPR. Evidently the 1990s had made him a full legend. After passing my oral interview of past diving experience, I signed up for the courses, bought Gary's book, and went home to study up on what it was all about.

I had read the term "technical diving" in magazines and even heard it used once or twice in a dive shop, but it was not fully clear to me what it actually meant. During my reading I discovered that diving was generally referred to as technical if it went beyond the artificial 130 foot depth limit of recreational diving, involved mandatory decompression stops, or involved penetrating shipwrecks or caves. In this respect, the scuba I had been engaged in over the years was indeed technical diving, and strictly speaking, I suppose the case could be made that this was true, but there was a difference.

"Technical" stood for both "technique" and "technology." Divers were now reaching greater depths through the use of advanced technique and technology that were new to the sport diving industry, but been used by the military and commercial operations for decades. I had the advantage of a fifteen year hiatus from this type of hardcore diving to help gain perspective, and the information I gleaned from reading Gary Gentile's book - and others to follow - had the markings more of revolution than evolution.

The aspect of technical diving that was immediately and dramatically innovative to me was the emphasis on a systematic, professional approach to the "extreme" sport. Information was not husbanded, but was widely disseminated and eagerly devoured by the "tekkies." Technique was always carefully espoused with the caveat to make your own final decision on how to set up your gear and ultimately on how to conduct your dive. But procedure and preferences were amazingly standardized. There even existed a concerted effort - although I was curious to see how it panned out in practice - to make the sport less macho, to put less pressure on not backing down or showing a weakness. I was to hear over and over again during my training that anyone could back out, could "call the dive," with no questions asked.

Strong elements of these influences always existed in wreck diving, but they were carefully stored beneath a thin veneer of machismo. But, on reflection, "macho" was the wrong word. There were always women deeply involved in the sport, and from what I could tell even twenty years ago, as a practical matter, there was little overt sexism. That is assuming that the individual - any individual - could handle the challenge they had chosen. And many women clearly could. Janet Beiser had always been Steve Bielenda's sole full time crew on the *Wahoo*. She would heft her doubles onto her back faster than most of the guys before plunging overboard with casual disregard to set the hook. Divers like her and Sally Wahrmann, Mary Artali, and Ceil Connolly routinely took on greater challenges than many of the other crew on the *Wahoo* and had dove the *Andrea Doria* well before it became "fashionable." They were certainly more accomplished divers then almost all of the *Wahoo* passengers.

SALLY WAHRMANN AND JANET BEISER. FROM THE AUTHOR'S COLLECTION, PHOTOGRAPHER UNKNOWN (1983).

Being a competent wreck diver and acting somewhat professionally in the haphazard definition as it pertained to sport diving at the time was the important thing; being able to "walk the walk." In 1982 anybody might be considered a "pussy," man or woman. This was not because they held back or were cautious, but because they talked a good game, but could not deliver. There was a remarkable sensitivity to avoid pushing other divers beyond their limits, but this was not guidance written in a manual or specifically addressed in an advanced course. There were no advanced recreational courses teaching deep diving (defined as greater than 130 feet), or how to properly decompress, never mind formal instruction in the use of mixed gases. Even discussing solo diving was taboo in most dive industry circles. Bits and pieces of military and commercial dive training filtered into the recreational sport, but learning from the experiences of other wreck divers was the only real classroom.

Technical diving had taken the most positive tenets of experienced wreck and cave divers out of the shadows and into the sunlight. The process had instilled a semblance of professionalism into a sport that previously had been ad hoc and extremely parochial. I recognized a pleasant irony in the explosive growth in the promotion of this relatively new technology and advanced instruction.

The dive industry stressed the need to self-police in the 1970s and early 1980s, to avoid teaching or even discussing potentially risky techniques; under no circumstances should any sort of inherently dangerous diving be encouraged. The logic behind the reasoning focused on a fear of the forced regulation that dominated civilian flying in the United States, which made it extremely expensive and difficult for the average guy or gal to participate. While working at the Island Dive Shop from 1979 to 1982, I had learned that the instructional emphasis was to stress the safety of the sport; we were not to talk to the average customer about the "other" type of diving, the deep wrecks, that was growing in popularity despite dive industry efforts. The dive industry wanted the plumbers and truckers to take up scuba, not just the few rich doctors who could afford flying lessons, and to keep the cost down meant keeping the regulation out. Decompression and deep diving were seen as too dangerous, as activities that would inevitably lead to high profile injuries and deaths that would invite future government restriction. Onerous regulation would translate into training costs too high to bear for a person of average means. The instructional agencies viewed it as an implicit sanction if they even officially acknowledged the existence of such practices.

The dive business was successful in that a net of regulation did not fall over recreational scuba as it had with private pilots. But the positive aspects of regulation imposed on the civilian flying industry were also missing: the open discussion of controversial practices on a national level, standardized, professional instruction, and ready access to information and equipment that might save one's life.

Due to the above-board actions of a few divers, primarily from the cave diving community, this all changed in the late 1980s and early 1990s. Sheck Exley, world-renowned cave diver, taught the first recreational trimix - a variable mixture of oxygen, nitrogen, and helium - course in 1987. By 1989 Billy Deans, a Key West dive operator, had introduced the advanced training to those diving in the open water and on wrecks. Billy Deans led a Florida group on a *Wahoo* expedition to the *Andrea Doria* in 1985, where he found out firsthand what a difference cold, murky water made while diving a wreck. Unfortunately, not all on that 1985 *Wahoo* charter would survive to benefit from the experience.

In the early 1990s Tom Mount formed the International Association of Nitrox and Technical Divers, and Brett Gilliam founded Technical Divers International. Others followed suit. What I viewed in the year 2000 was a relatively mature, fully evolved, and standardized instructional process that appeared to have all the advantages of the flying industry while still managing to avoid the stifling, costly cloud of wide-spread government regulation.

The scuba diving business would always be parochial, but the competitive nature of the industry actually moved the sluggish behemoth of prevailing attitude toward greater openness. Pandora was out of her box, and the industry might as well make the most of it by selling safe and tried gear, offering professional instruction, and getting more people involved in the sport.

If anything, technical diving was perhaps too accessible. The required experience level to qualify for some of the new types of advanced courses appeared to be minimal, but perhaps the sport was in a mid-cycle swing and was about to correct toward more balance. I suspected that the inevitable accidents would gradually tighten up the self-policed restrictions of the various instructional agencies; the threat of the business-killing costs of federal regulation must not have entirely disappeared.

Overall, it was impressive. I never thought the insular, protective nature of recreational scuba would allow the open teaching of technical type courses under any circumstances. On the other hand, it had taken over thirty years.

Sport divers "broke the rules," however, even before the "rules" of mass-market instruction existed in writing. The day after the *Andrea Doria* sank, on 27 July 1956, adventurers Peter Gimbel and Joseph Fox went to extraordinary lengths to charter a fishing boat to take them out to the shipwreck. With their primitive wetsuits and double-hosed regulators they followed the air bubbles still streaming from the sunken ship until they reached the liner. A photograph of the two "technical divers" was printed in Life magazine less than two weeks after the disaster. The allure of the *Andrea Doria* inspired Peter Gimbel with a life-long passion for the shipwreck, and he would return to the *Doria* consistently over the course of twenty-five years, twice with major commercial salvage operations. His near obsession focused on the mysteries of the *Andrea Doria's* sinking and what had happened to the valuables secured in two safes.

Why had the *Andrea Doria* taken on such a dramatic list to starboard so quickly after being sliced open by the *Stockholm*? Why did the *Doria* flood so rapidly, when her design was supposed to maintain watertight integrity, to keep the 697 foot liner from lurching over at more than a fifteen degree angle? Rumors of a treasure, a Purser's safe brimming with valuables, grew with time and took on a life of its own. The amount of reputed riches seemed to increase with each passing year.

BREAKING THE RULES

But for Gimbel it had to have been more than a quest for specific answers, and he was certainly not inspired by money. Peter Gimbel first demonstrated the *Andrea Doria's* irresistible attraction to divers, but like Sir Edmund Hillary on Mount Everest, his was but the first success of many in the pursuit of a challenge that quickly became contagious. The vast majority of adventurers who visited the *Andrea Doria* over almost half a century spent far more in money preparing for their dives than any artifacts they recovered were actually worth.

The first major enterprise to return a significant piece of the *Andrea Doria* to the surface began in 1963 and lasted through 1964. Two wealthy entrepreneur/adventurers converted a 125 foot surplus Coast Guard cutter into a salvage vessel renamed the *"Top Cat."* The boat was replete with heavy salvage equipment, a diving bell, and a recompression chamber for emergency use to save a bent diver. The original purpose of the *Top Cat's* several expeditions to the wreck included an exploration of commercial salvage possibilities and an eventually failed photo opportunity for Life Magazine. Finally, the determined group - with the aid of four divers on leave from the Navy - embarked on salvaging the most ambitious piece of the *Andrea Doria* that recreational divers could hope to achieve without the full support of a commercial or military operation: the life sized bronze statue of the wreck's namesake, Admiral Andrea Doria.

The operation took eight days of intensive labor. Explosives were used to blow a five by eight foot hole through the bulkhead that separated the Promenade Deck from the First Class Lounge where the good Admiral still stood horizontally on a pedestal anchored to the deck. Pairs of divers took turns hack sawing the legs of the 750 pound statue, arduous work at 210 feet of depth inside the wreck.

The *Top Cat* divers were successful, and the statue's surfacing into the sunlight signaled the end of the expedition. The excitement generated from the amazing feat of scuba salvage soon waned, and Admiral Andrea Doria found a home in a Pompano Beach, Florida, hotel owned by one the *Top Cat* expedition's financial backers. Eventually the *Top Cat's* Captain, Dan Turner, would gain possession of the statue. In his last known location,

SETTING THE HOOK

Admiral Doria was standing proudly in Captain Turner's front yard (sans feet) in Fernandina Beach, Florida.

The *Andrea Doria* expeditions for the first several decades after sinking centered on modest salvage investigations and movie making enterprises. Divers visited the wreck using "heliox," a mixture of helium and oxygen, for the first time in 1968. By avoiding the use of nitrogen in the breathing mixture, a clear head could be maintained at the depths required to dive the *Doria*.

Eliminating the dangers of nitrogen narcosis made the underwater work of the various photographic enterprises safer and theoretically more productive. But it required a great deal of surface support and financial backing to get the divers in place for their shoots. This all took time, and the longer and more involved the expedition, the more problems that could be expected from the greatest intangible to planning - the weather. Most of the handful of ambitious undertakings to document the *Andrea Doria* on film, or to retrieve her most valued possession, the Purser's safe, met with varying degrees of failure due to the complex requirements of the projects and Mother Nature's fickle hand.

In contrast, as one would expect, the first expeditions of limited scope proved more successful, largely due to the simple, eminently attainable goals of the small-scale projects, starting with Peter Gimbel's first underwater photographs the day after the *Doria* sank. In 1966, a few sporadic, hearty scuba divers began chartering fishing vessels to take them to the wreck site. To simply touch the *Andrea Doria* ordinarily qualified the trip as a success of sorts, but to return with a recognizable piece of the Grand Dame of the Italia Line was the intrepid weekend warrior's goal. In 1967, veteran diver John Dudas returned from a three-day charter with the prize of a lifetime - the ship's compass and the binnacle cover that housed it. His future wife, Evelyn, had joined in the expedition, and she became the first woman to dive the *Doria*. John Dudas unfortunately died on a deep northeast wreck twenty years after his *Andrea Doria* charter.

There were not many sufficiently capable sport divers who were willing to risk the brief visits to the sunken liner. Access to the *Andrea Doria's* interior was severely limited, and the most impressive artifact that a diver could reasonably expect to retrieve in the 1960s and 1970s was a brass framed Promenade Deck window, still no small prize. One commercial operation did manage in 1973, however, to cut a small hole in the massive, Foyer Deck doors in an unsuccessful search for the elusive Purser's safe. Very few scuba divers, Gary Gentile being an exception in 1974, risked venturing through the narrow hole into the abyss of spider-webbed cables and debris. Fifteen to twenty-five minutes was all that was allowed a scuba diver breathing air out of double, eighty cubic foot tanks, which was a negligible amount of time for a diver to get free if they became entangled while sifting through the wreckage in the blackness of the ship's interior. There was only one way in, and one exceptionally tiny way out.

In 1975 Peter Gimbel enlisted the help of a commercial firm, International Underwater Contractors, to film a movie, one that would hopefully reveal the secret of the *Andrea Doria's* sinking. The company's commercial divers assisted Gimbel and his photographers throughout the expedition. They breathed "trimix," a calculated proportion of helium, nitrogen, and oxygen, supplied to the diver's via lengthy hoses from the surface. The advantage of trimix was the alleviation of the narcotic affects of nitrogen without having to breathe pure helium and oxygen. Helium made the divers colder due to its higher thermal conductivity and required a slightly longer in water decompression. Helium was also expensive. Trimix was a compromise between the positive and negative characteristics of both helium and nitrogen.

During decompression, the support ship would pump down pure, 100% oxygen to the divers to lessen the time required to hang. This was done at the maximum depth thought safe, which was evidently not safe enough; Gimbel at one point underwent the uncontrollable convulsions of oxygen toxicity while decompressing. If not for the fact that he wore a full-faced mask, one that did not require him to keep a regulator in his mouth, he would have likely drowned.

Gimbel's 1975 expedition was successful in that it accomplished far more than any other to date. What it could not determine, however, was the precise cause of the *Andrea Doria's* extreme initial list. The exact point where the *Stockholm's* bow had penetrated the *Doria* could not be accessed due to a mountain of fallen rubble inside the now sideways Foyer Deck. Gimbel was able to see far enough through the rubble to lead him to suspect that the absence of a watertight door to the ship's generator room had caused the excessive flooding, but he did not know for certain.

In 1981 he would find out.

CHAPTER THIRTEEN

Beckoning from the deep

Watching for sharks and passing ships, swimming with whales, fishing, and photography - the recreation at the surface above the *Andrea Doria* was worth a *Wahoo* charter in itself. It should not be surprising that more than a few people over the years paid for passage on a *Wahoo* charter with no intention of diving at all. Despite the super tankers and giant freighters that steamed by at times uncomfortably close, the sense of isolation anchored over the *Andrea Doria* could be profound. But in subtle contrast, sitting onboard the fiberglass *Wahoo*, an out of place manmade platform in the midst of nature, there was also an awareness of being surrounded by watching eyes.

Once the occasional morning fog burned off under the warming sun, the blue sky's merge with the infinite ocean inspired a sense of welcome vulnerability under nature's benevolent power. A balanced blending of thrill and serenity came easily those mornings, putting one at peace on the calm sea under perfect weather. These were the best days; a light wind to disperse the fog and temper the hot sun, flat sea and quiet air leaving the senses searching for sound, and the low bustle of growing activity below decks as divers woke reassured in the knowledge that the welcome solitude would only last just long enough.

Finback Whales, graceful in their giant flow, would pretend to be shepherded by the *Wahoo* nearing dive site. After the commotion of anchoring

SETTING THE HOOK

had subsided the whales often returned, slowly arcing effortlessly just below the surface in their approach. The huge mammals were likely as curious as their human cousins, and the Finbacks would almost allow a swimmer to touch them. A swimmer could come close, but never quite manage that last twenty feet, particularly if the whale in question was a mother herding her baby just below the waterline. The whales seemed to prefer it nearer the surface when paralleling the *Wahoo's* path. One had to wonder if they were aware of man's permanent mark on the spot, of the nearly 700 foot shipwreck, that lay below their bellies. It is seductive to subscribe to the feeling that such a magnificent sea creature is omniscient and completely aware of the most trivial circumstance unfolding in its surroundings. Just as a person can maintain a near total awareness of the movement of others in a solitary room, it seemed that the Finbacks must be able to effortlessly recognize all life around them for miles.

But even if the whales were aware of the shipwreck below, did the Finbacks attach significance to the *Andrea Doria*, or was it simply another pile of ocean rubble? Were we playing the part of the ugly American tourist in a foreign land, visitors to be toyed with, humored, but only accorded respect because of the ultimate devastation we were capable of inflicting? Looking out across the endless ocean vista and hearing the gentle lap of waves at the waterline, nature felt in balance and anything seemed possible.

The sharks did not appear to be nearly as intuitive as the Finbacks. The grapnel hook rarely had the opportunity to snag the *Andrea Doria* before the first eerie dorsal broke the surface. Sharks scared me, regardless of how much I read about their more likely true nature, or relied on the ancient advice of mothers the world over, "If you don't bother them, they won't bother you." It was something about being completely at their mercy, awkwardly swimming in the water enveloped in the cumbersome dive gear necessary to keep us alive. But then again, it was even more uncomfortable for me in nothing but a swimsuit in the middle of the sharks.

The smallest of sharks could wreak havoc if they so desired, and one didn't need to witness the violence for it to be instinctually understood. A single glance at the sleek, bullet forms gliding through the water, navigating effortlessly with imperceptible flicks of their fins was enough for me. You were theirs if they wanted you. The indisputable fact that they almost never, ever, wanted to eat humans did not mitigate the nagging fact that they could. The presence of sharks was never enough to stop me from diving, and once in the water I would usually quickly forget about them. But standing on the deck of the *Wahoo* could afford me too much time to think about the unlikely possibilities. "A mind is a terrible thing," as half the saying goes.

One hundred miles offshore there were a wide variety of shark species, but by far the most prevalent were the Blues. Blue sharks do not have a particular reputation for ferocity, but the commonly held view in the early 1980s was that they could distinguish certain colors, and had a marked preference for "yum-yum yellow." I do not recall whether it was a Cousteau TV special or maybe a National Geographic article that started the iron clad belief in divers that Blue sharks got aggressive around the color yellow. As far as I was concerned, it was true. Yellow had always been a traditional color for aluminum scuba tanks. Funny how that works, isn't it?

Sally Wahrmann probably did not find it overly funny, at least for a few minutes after one *Doria* dive while hanging on the anchor line decompressing. Sally shared a home with Mary Artali and the two were regular *Wahoo* crew. They were both large women who occasionally would diet back their size - Steve once bragged that they had "lost a person between the two of them" - but they were exceedingly comfortable with their identities: amazing divers, interesting people, and loyal and faithful friends. Sally was an accountant who taught the same at the university level. A licensed CPA, she was comfortable "telling it like it is" to anyone at pretty much any time, enough so that on the *Wahoo* the letters "CPA" came to stand good naturedly for "Certified Pain in the Ass." On overnight trips Sally would double as the *Wahoo's* cook, but she and Mary acted primarily as Dive Masters; certified individuals tasked with the responsibility of verifying the proper certification credentials of passengers and ensuring the safe running of the dive operation topside. We all loved Sally and Mary, but that did not mean they got off the hook from pranks.

SETTING THE HOOK

With about twenty minutes remaining in her decompression before she could surface, Sally felt a sharp impact at her back. The blow momentarily pushed her off the anchor line and she fought the current until managing to grab back a hand hold. She took a quick look around and saw that an eight foot Blue shark was aiming its pointy snout in her direction for a second run. The shark continued making passes, avoiding further physical contact, but clearly keeping Sally's attention. The decompression must have gone by particularly slowly with her finned-friend keeping company during the hang. In retrospect, the rest of the crew topside could have considered waiting for Sally to finish her dive before tossing "chum" in the water to lure sharks for rod and reel fishing, but she was tough and we figured that she could handle a wimpy shark. Fortunately for us, Sally was one hell of a good sport.

It was normal to have the company of one or two Blue sharks when hanging on the anchor line. They must have frequented the wreck as well - that was where most of the fish were - but I never saw one on the hull of the *Doria*. Most of the fish that formed the mini-ecosystem under the keel of the *Wahoo* were tropical strays, fish that had been caught up at the far edge of the north flowing Gulf Stream which came closer to shore in the month of July. The surface water temperature was erratic, ranging from fifty to sixty degrees, depending on which way the currents were mixing the warm Gulf Stream, the Labrador Current from the north, or the colder waters from 240 feet below. It was more than a touch surreal to dive off the coast of New York and watch tropical fish swim by.

The mountain of commercial nylon fishing nets that draped the "side" of the *Andrea Doria* (actually her deck and what was left of the superstructure as the ship lay on the bottom) never rotted away and year after year continued to make their catch. But the massive hull of the *Doria* more than compensated nature by acting as an artificial reef, and the wreck provided a rich breeding ground for the marine critters that otherwise could not survive in the barren sand. Fifty pound Pollock and Cod provided fishing entertainment for both the humans above on the *Wahoo* and the sharks below. In addition to the Blue sharks, I was to see Makos, Tigers, and once a huge lone Thresher shark. Thresher sharks are clearly recognizable by an extremely tall tail fin.

EERIE VIEW LOOKING THROUGH SNAGGED FISHING NETS ON THE PORT BOAT DECK RAIL.
PHOTO COURTESY OF BRADLEY SHEARD.

One sunny afternoon while standing on deck watching the never ending entertainment that emanated from the depths, the massive tail of a Thresher shark broke the surface only thirty yards from the *Wahoo*. This shark was big - probably twelve feet long - and unlike the narrow Blues, it was extremely wide in girth. The tall tail fin sliced lazily through the water, hesitated in the *Wahoo's* vicinity for a moment, and then the shark began to slowly swim away. Gary Gilligan and Hook jumped in the inflatable chase boat with cameras in hand and motored out for a closer look. Evidently the sound of the outboard engine was irritating the Thresher shark, and the big fish flicked its powerful tail occasionally with a bit more force to stay ever so slightly beyond the reach of the intrepid photographers.

We watched from the deck of the *Wahoo* as the inflatable motored further and further into the distance. Repeatedly, the chase boat would close the gap to the Thresher until the giant tail swished effortlessly and the shark would pull away. Gary and Hook, finally losing patience, gunned the motor in an attempt to catch up. At about a mile away, and now unable aboard the *Wahoo* to distinguish between Gary, Hook, and the pontoons on the inflatable, the Thresher's giant tail could still be clearly seen. Suddenly, the tail snapped back directly at the inflatable and the water erupted in a geyser of white spray. For a heart-stopping second I could see the bottom of the inflatable. Not a good place to be, about to be capsized next to a clearly aggravated Thresher shark. The inflatable barely managed to stay upright, and Gary and Hook turned and sheepishly motored back to the *Wahoo*. They claimed that they "had gotten bored" with the chase. Scratch one photo opportunity.

Big Artie Kirchner was six-foot six-inches and 260 pounds, which made him a strange candidate for a unique method of shark photography. While several buddies grabbed his legs, he held his breath, and was lowered headfirst off the dive platform at the *Wahoo's* stern holding his camera at the ready. Several sharks had been swimming through the residual chum field from earlier fishing efforts, and Artie thought it would be a good way to shoot a picture of one of the Blues. I watched and reveled temporarily in the comfort of knowing that others in the group did stupid things as well. It took only one yell of "Mako!" for Big Artie, perhaps not with the grace of a ballerina, but most definitely the speed, to reverse himself out of the water and flop onto the deck. Mako sharks were decidedly more aggressive than Blues.

Once the divers were out of their dry suits (or in my case, "variable volume wetsuit," as Steve and Sally liked to quip) with their gear put away, it could be enticingly easy to forget for a moment where the *Wahoo's* anchor line was secured. The presence of the *Andrea Doria* rarely made herself known to those on the surface with the great divide of the depth almost completely separating the two worlds, but I remember a notable exception.

On a calm glassy-water day, twenty-eight years after the disaster that claimed the *Andrea Doria*, single drops of fuel oil dripping in reverse from the bottom of the ocean saw the sunlight once again. The steady, even rhythm of each slowly surfacing drop from the *Doria's* fuel tanks, the tanks that had been nearly empty and had exacerbated her ungainly list after the collision, marked the passage of time without urgency or concern for the outside world. The cascading rainbow of color, finally able to disperse and evaporate, to escape the shroud of steel that had trapped it for so long, beckoned to the watchers on the surface. Only the complete absence of wind allowed for the brief glimpse of the past on the ripple free ocean. I had a nearly irresistible urge to try and follow the oil downward, to discover its exact source. It was a persistent and patient message, a Morse code of colorful blooms repeated over and over. That the *Andrea Doria* had a message seemed at times indisputable, but the nature of her tidings was anything but clear.

Peter Gimbel, more so than any, was obsessed with deciphering the great liner's message. After two and a half decades of attempts of varying complexity, Gimbel put forth a final grandly successful effort. The lessons learned from previous efforts ensured the expedition's success by combining daring, advanced technology, careful planning, and - most importantly - money. The project that Peter Gimbel and his wife, Elga Anderson, organized in July of 1981 would cost one and a half million dollars and took full advantage of the commercial diving assets of the Oceaneering International Corporation and their salvage vessel the "*Sea Level Two.*"

With thirty-one divers and crewmembers dedicated to the project, Gimbel's team set out with three specific objectives: to retrieve the *Andrea Doria's* Bank and Purser's safes, to discover the ultimate cause of the rapid

SETTING THE HOOK

flooding that spelled the doom of the ship, and to capture the entire odyssey on film for a movie. Gimbel enlisted the aid of veteran underwater photographers Jack McKenney and Bob Hollis for the last goal. A team of Oceaneering's experienced commercial divers, led by Ted Hess, would do the lion's share of the physical work, leaving the project management to Gimbel, Anderson, and the professional support crew onboard the *Sea Level Two*.

The *Sea Level Two* was equipped with an onboard recompression chamber and a four-ton diving bell, both necessary for the type of prolonged exposure that the divers were intending. Ted Hess and his divers would be put into "saturation," compressed on deck to the point where the gas mixture of helium and oxygen they were breathing became completely saturated in their bodies. The great advantage of saturation diving was that the underwater workers need not decompress after each individual dive, but instead stayed at the wreck's pressure for essentially as long as they desired. Hot water was pumped down to the diver's suits to keep them warm, and surface supplied gas gave them a virtually unlimited amount of mixture to breath.

When the divers were finished with their work for the day on the *Andrea Doria*, they would return to the bell which was shackled by a steel cable to the wreck's hull. They would then be raised to the surface and transferred to the recompression chamber on the boat deck through an airlock. The pressure required to keep the helium and oxygen mixture in saturation was maintained at a constant level whether the divers were on the wreck, in the bell, or on the deck of the *Sea Level Two* in the recompression chamber. Food and drink was pressurized in the recompression chamber's trunk and removed and eaten by the divers while still at the wreck's ambient pressure. With virtually unlimited time on the bottom, the work would presumably go quickly and allow access deep into the *Andrea Doria's* interior.

The payback owed by the divers for this freedom came at the end of their stay in saturation, when it would take two full days to decompress and exit the confines of the chamber. And, of course, there were a few nagging concerns. If the chamber blew out a seal and de-pressurized all those within would quickly die. To visit the *Andrea Doria* meant doing so on her terms. The divers would be prisoners to the wreck, the bell, or the chamber for the duration of their saturation.

The first obstacle facing the expedition was ready access to the *Andrea Doria's* interior. The two sets of Foyer Deck doors exposed on the *Doria's* port side were the obvious target. Once removed, the doors would allow the degree of access required to get to the parts of the ship needed to accomplish Gimbel's goals. The project was immediately plagued with difficulty. The diving bell made its first descent on August 3rd, only to be brought back up after little could be accomplished due to inclement weather. The *Sea Level Two* was on a four point mooring system. Four large anchors bracketed the *Andrea Doria*, enabling the *Sea Level Two* to maintain precise position over the wreck by adjusting the slack among the four anchor cables. If the seas got rough and it became too dangerous to raise and lower the bell, the divers would have to sit out the bad weather in the recompression chamber.

The main down line - the steel cable guideline between the diving bell and the *Andrea Doria* - tore loose the second day on the wreck. In the weeks to follow, the down line would be found several times mysteriously unshackled, not broken, but simply unbolted from the wreck, leaving the crew to speculate that perhaps it was a message from the ghost of the great liner. On August 7th, Ted Hess finally finished torching off the Foyer Deck doors. It had taken twice the project's allotted time.

With the Foyer doors removed, the divers were able to enter the *Andrea Doria*, search for the safes, and if found, even have sufficient room at the cut-out in the hull to remove the two-ton steel boxes from the wreck. But there were other challenges to face first. Ted Hess descended into the Foyer Deck, past the corridor at 205 feet leading to the First Class Dining Room and at 220 feet came across two huge, completely intact glass windows. Thinking he had found the Bank, he smashed the panes, worked his way inside, and discovered to his dismay that he was in the ship's gift shop. He continued further down into the Foyer Deck.

The effects of gravity and years of storms and currents had taken a toll. Everything that was not bolted down, and many things that had been bolted down, had shifted to the *Andrea Doria's* starboard side, now flush with the sand, and created a huge debris pile. After ten days of sifting through the rubble and experimenting with various methods of removing the debris from the interior of the ship, the divers found a metal safe secured tightly to the bulkhead. Peter Gimbel decided it was time to join the team underwater, and he began to saturate the helium into his body under the pressure of the recompression chamber on the *Sea Level Two's* deck.

As luck would have it, no sooner had Gimbel become a prisoner of the chamber than hurricane Dennis traveled north toward the *Sea Level Two*. With mounting frustration the group of divers huddled in the chamber, bouncing in the building waves, waiting for the storm to pass. It was not until five days after the discovery that the divers were able to reach the Bank safe and cut it loose.

Peter Gimbel's decision to enter saturation may have had less to do with the Bank safe, however, than his desire to complete the second goal of the expedition - to explore deep within the *Andrea Doria* and discover the secret of her rapid flooding. For decades, rumors had abounded that the cause of the *Doria's* uncontrollable flooding was due to a missing watertight door leading to the ship's generator room. Once the generators became swamped, the *Doria's* pumps would be unpowered and her fate sealed. If the missing door were the answer, the fault for the *Andrea Doria's* sinking would clearly lie at the feet of her crew and the Italia Line. Peter Gimbel was determined to learn the answer.

Ted Hess and Peter Gimbel squirmed through a ventilation shaft to the generator room, only to be confronted by a steel grate between them and the spot where the door should have been secured shut. Stymied for a time, Ted Hess finally searched out a narrow ladder-way leading one level down, and then back up to the passage that they wanted to traverse. When the pair finished swimming down the cramped, darkened passage, they were confronted with far more open space then they had anticipated; the divers found themselves literally on the outside of the *Andrea Doria's* hull and in the open sand.

It took a minute to absorb the meaning of the find, but it soon became clear. The collision with the *Stockholm* had rendered a far larger hole in the side of the *Andrea Doria* than anyone had expected. The jagged tear in the *Doria's* side was about eighty feet long, which was roughly twice the length estimated by most experts. Whether the generator room door had been in place or not was moot; the damage from the collision was so extensive that the *Andrea Doria* was doomed from the moment of impact.

The discovery was decisive; after twenty-five years Peter Gimbel had settled the debate surrounding the *Andrea Doria's* sinking. All in the project were elated. On top of that, while it had taken nearly a month, sufficient camera footage had been compiled to finish the documentary. All that remained was locating and recovering the Purser's safe.

The perception of victory or defeat in any great challenge is directly related to expectations, to what criteria ultimately define success for an individual. The *Andrea Doria* seemed eager to suck one in, always tempting a diver to take a step too far. Whether it was a twenty minute adventure to the wreck or an elaborate, commercially supported operation, the temptation and accompanying risk were always there. I felt it - and fought it - by not exploring further in search of the ship's Gift Shop after realizing that I had no clear idea of which way to go. But it was tough. After seeing the trinkets of costume jewelry recovered by Gary Gentile, it was really, really tough to stop looking. And when I had not fought the urge to go a step further, I almost died after getting lost in the First Class Dining Room.

Peter Gimbel was not looking for anything as mundane or monetarily worthless as costume jewelry. But when all was said and done, whether looking for a safe, jewelry, china, or simply trying to touch the wreck, if personal limitations were not respected, then catastrophe became likely. Gimbel and the other saturation divers were sick with flu like symptoms and fever from the extended time spent under pressure. But how could they stop now, just short of total victory, how could they, "Call the dive"?

It took the gentle - and not so gentle - prodding of Elga Anderson and the top-side crew of the *Sea Level Two*, but, finally, the divers came to a disciplined understanding that the project was teetering between an already attained triumph and a hubris inspired tragedy. The Purser's safe was left on the *Andrea Doria*, lying somewhere beneath the rubble pile in the

SETTING THE HOOK

Foyer Deck. The *Sea Level Two* pulled its four anchors and turned toward Montauk, New York.

Two days later, the divers emerged from the recompression chamber to the chatter of news cameras onboard the salvage vessel, now pier-side at Montauk, New York. The Bank treasure was sealed and transported away for "safe" keeping, eventually to be opened live during the premier showing of Gimbel's documentary film aired several years later. The deteriorated and waterlogged bills turned out to be worth more as auctioned souvenirs than their face value. But divers around the world took excited notice of the other treasures unveiled at the Montauk docks - fabulously colorful hand painted china - a wreck diver's dream. It was laid out for all to see and there was plenty more inside the *Andrea Doria*, waiting for anyone with the daring and skill to dive into the wreck and pick it up.

The seed had been planted. The quest began.

CHAPTER FOURTEEN

"Block-head" Island

Two weeks before Peter Gimbel's expedition rigged the main down line to the *Andrea Doria*, the *Wahoo* had been grapneled into the exact same spot. The finishing touches had been put on the custom built *Wahoo* only months earlier, and the summer of 1981 was Steve Bielenda's first opportunity to explore further from shore than his previous boat, only thirty-two feet long and also named the *Wahoo*, was capable of safely venturing. The *Wahoo's* first journey to the *Andrea Doria* was organized to commemorate the twenty-fifth anniversary of the liner's sinking.

Despite six to eight foot seas that persisted for the duration of the expedition, Steve, Hank Keatts, and Billy Campbell managed to set a plaque on the wreck to commemorate the event. Few others onboard the *Wahoo* attempted to dive in the heavy seas, and most passengers stayed close to the rolling rail, victims of unrelenting sea sickness. One exception was a passenger named Marty German, who ended up getting bent for his efforts and had to be evacuated by Coast Guard helicopter. He recovered completely from the incident, only to die several years later on a deep dive to a different wreck.

This first *Wahoo* charter to the *Andrea Doria* happened to hook into the Promenade Deck almost directly on top of the two sets of doors that still covered the entrance to the Foyer Deck. None aboard the *Wahoo* knew it at the time, but they were in the exact spot that would prove to be the preferred tie-in location for many years to come once Gimbel's team removed the Foyer Deck doors.

For most on the charter, Steve Bielenda and Hank Keatts included, this was the first trip out to the famous wreck. Rapidly building storms gave little warning and a boat could endure quite a pummeling during the five hour motor to safety at the nearest port on Nantucket Island. Eight foot waves were no threat to the *Wahoo*, and in a pinch hearty divers could take a stab at negotiating the pitching deck to get in and out of the water, but eight foot waves made for miserable diving conditions and traveling to and from port was not much better.

But at least the new, second "*Wahoo*" could handle the rough off shore waters more responsibly than Steve's original boat. Steve was running a business and the only way that he could make money even on a pristine day this far from shore was by carrying more passengers. Fuel expenses incurred crossing the ten hours of open-ocean between the *Andrea Doria* and Montauk (fifteen hours if steaming directly to the Captree Boat Basin) required more paying passengers than a local charter in order to turn a profit, and the original *Wahoo* could not accommodate enough divers to make it financially viable. Steve could now run profitable expeditions to the *Andrea Doria* in addition to reaching out to a whole new breed of off-shore wrecks only previously visited haphazardly or not explored at all.

The *Andrea Doria* was first on Steve Bielenda's list of unfamiliar ship-wrecks to explore. She had the name recognition and historic background to fill a charter. Besides, Steve wanted to touch the liner for himself. The *Wahoo* did not stay over the *Andrea Doria* long in 1981, and shortly after placing the plaque on the wreck's hull for a photograph the unrelenting seas finally beat the intrepid group into submission. Not wanting to risk putting in a diver to free the hook, Steve cut the yellow, expendable polypropylene line tied into the wreck and left the anchor attached to the *Andrea Doria*. The *Wahoo* raced the weather back to refuge at Block Island, Rhode Island. Two weeks later the *Sea Level Two*, moored by its four giant anchors directly over the *Andrea Doria*, would see a yellow line flapping on the surface. Project photographer Jack McKenney followed the rope to the

"BLOCK-HEAD" ISLAND

Promenade Deck to secure the shackle for the main down-line. They never knew whom to thank for the guideline to the wreck.

While Peter Gimbel's commercially supported *Andrea Doria* expedition was underway, the *Wahoo* was still nearby and within VHF radio range, now searching for the passenger liner *"Republic,"* another victim of a fog bound collision in the well traveled shipping lanes leading to New York. The White Star Liner *Republic* was rammed by the Italian ship *"Florida"* in 1909 after the *Florida* became lost in the fog and strayed thirty miles off course. A total of six crewmembers died in the mishap, four on the *Florida* and two on the *Republic*. It took thirty-nine hours for the *Republic* to sink, during which time hundreds of *Republic* passengers and crew were transferred first to the *Florida* and then from the *Florida* to other more sea worthy rescue ships as they arrived on the scene. A total of 2,494 people were therefore transferred at sea without the loss of life, but only because 822 *Republic* passengers were evacuated twice. Still, the ship to ship transfer of this many passengers without a fatality set a maritime record. The rescue was also noteworthy for being the first successful use of wireless in a disaster at sea. The story of the *Republic* stood in interesting contrast to the *Andrea Doria* and *Stockholm's* failure to successfully utilize a relatively new technology, radar, to stave off tragedy. The *Republic* also carried valuable cargo including - probably - a very large quantity of gold, which made her all the more intriguing to both professional and amateur would-be salvors.

The *Wahoo* had been chartered by a Martha's Vineyard dive store owner, Marty Bayerle, to find the *Republic*. Bayerle brought with him a group of "professional" commercial divers who were, oddly enough, unfamiliar with deep-ocean wreck diving operations. The commercial divers had not worked deeper than one hundred feet before and were accustomed to operating with unlimited, surface supplied air. This time they would use double tanks to depths that were certainly greater than one hundred feet, but how much deeper no one yet knew.

For two days the *Wahoo* scanned the ocean with a towed sonar array, patch-working the vast, open nothingness in a search pattern grid. Even a large wreck in the open ocean is but a speck on the bottom without reliable information on its precise location, and it did not take much of a miss on the sonar to come up empty handed. To make matters worse, the researched latitude and longitude coordinates of the *Republic's* purported position were suspect from the start. Fortunately, an instant of luck triumphed over weeks of possibly futile effort. On the third day Steve spied a large, orange buoy in the distance, very much out of place this far from shore. The *Wahoo* approached, and Steve's crew - Hank Keatts the professor and Janet Beiser - attempted to haul it in, but were unable to budge the marker. It was immediately apparent that the line was firmly secured to something very big.

Like many shipwrecks, the *Republic* had first been found by a fishing vessel that had entangled her expensive nets permanently on the underwater obstruction. It would have been prohibitively expensive to hire a diver to retrieve the lost fishing gear so deep and far from shore, but the trawler captain could, however, take measures to ensure that he did not snag a net on the obstruction again. Before releasing his hung up net, he tied off two large, orange buoys, one at each end of the hang, to use as a visual reference to avoid fishing in the immediate area in the future.

A quick scan of the sonar display corroborated the fact that the *Wahoo* was indeed over something very big, most likely in the 700 foot range, about the size of the *Republic*. But was it the *Republic*? There was only one way to find out, and the professional divers got suited up and descended down the anchor line. They made it to 190 feet, where in the excellent visibility they saw the outline of a debris field, but nothing else. In the discomfort of the unfamiliar diving gear and new experience of narcosis they ran out of air and barely made it back to the surface safely. The *Wahoo* was committed to another charter later in the week and was forced to leave the site at the end of the day, not allowing the "professional" divers another opportunity to try again.

Marty Bayerle asked Hank if he was willing to dive the wreck to solve the mystery. Hank agreed, and Janet quickly volunteered to go with him. Hank carried a Nikonos Three underwater camera to photograph any visible features of the wreck that might help in identifying the phantom ship.

Hank planned a dive to 200 feet deep - the point where the "professional" divers claimed to have seen the debris sticking up off the ocean floor - donned his twin eighty cubic foot tanks of air, and jumped over the side.

With Janet Beiser in tow, Hank descended down the anchor line and watched the rotation of the needle on his depth gauge, which only read to a maximum of 150 feet. Most depth gauges at that time did not indicate any further, who would actually dive that deep? There was no mechanical stop to keep the depth needle from moving past the 150 foot mark, however, and Hank could estimate his true depth by looking at the hand's second revolution around the dial. At approximately 200 feet, Hank saw the debris field below him, but far below him. He continued down. Once on the sandy bottom, with only a best guess at his true depth, Hank finally made out a huge, dark shape in the distance. Janet caught up, indicated via hand signals that she was narced, but wanted to continue on.

Hank was concerned about the depth. His planned decompression for a twenty minute dive to 200 feet was not going to be sufficient, and he had nothing better than a guess at his true maximum depth. He decided to cut the dive short to fifteen minutes, just enough time to swim to the darkened hulk and snap a few pictures. Slowly, the giant ship came into a semblance of focus, and Hank was able to see passenger loading doors and several rows of portholes. For some reason the Nikonos Three camera would not work and Hank, unable to take any photographs and out of time to explore further, returned with Janet to the anchor line and started up.

Using his best judgment and a good deal of Kentucky wind-age, Hank decided to double the decompression time at each required stop planned for the 200 foot dive. Once on the deck of the *Wahoo* and able to reference the boat's depth sounder, Hank and Janet realized they had actually descended to an incredible 260 feet while breathing from double eighties filled with air. Fortunately, Hank's best guess had been fairly accurate, and the Navy Decompression Tables indicated that they would probably not get bent.

Bayerle, frustrated as he may have been by the *Wahoo's* commitment to return to shore, had to be satisfied with locating and identifying the *Republic* for future divers to explore. Marty Bayerle and a group of qualified

SETTING THE HOOK

commercial divers did return several years later and retrieved multiple artifacts that they auctioned off, perhaps to help defray the cost of their expedition. Hank Keatts contacted the Nikonos Camera Corporation to determine why their product did not work at such a critical time. Nikonos wrote back that the pressure at 260 feet had been too great to allow the film to advance. Northeast wreck diving to 260 feet deep while breathing air from double eighty cubic foot tanks was not something that many people could do and return to the surface alive.

For the three days that the *Wahoo* searched for the *Republic*, the crew had entertained themselves by listening to the VHF radio communications between Peter Gimbel on the *Sea Level Two*, the *Doria* Project's provisioning support boat, and shore. Curiosity sparked, Steve decided to make a slight detour on the *Wahoo's* trip back to port to take a look at the elaborate project. Despite skirting the site of the *Andrea Doria* by five miles, Gimbel sent a boat out to chase them off. Evidently he was concerned about "claim jumping" of some kind on the part of the *Wahoo*. The lure of the *Andrea Doria* did strange things to people. The *Wahoo* headed back to port and Peter Gimbel apparently never lost his conviction that the *Wahoo* was up to some sort of conspiratorial plot.

Twelve months later, with barely one hundred dives under my weight belt, I was aboard the *Wahoo* for my first attempt at diving the *Andrea Doria*. The *Wahoo* fought the eight to twelve foot seas for ten hours, putting us at less than a dozen miles away from the *Andrea Doria* when Steve made the decision. The weather was not forecast to improve, and Steve was not going to allow anyone to attempt to dive in the dangerous conditions. In addition to being virtually assured of diving in a seasickness-induced state of extreme dehydration - a condition making one particularly susceptible to the bends - the rough water presented several other obstacles. The greatest challenge came with entering and exiting the water. With the *Wahoo's* gunnels high and dry one second, and twelve feet lower the next, balance and timing were critical. Screwing it up could mean being knocked unconscious on the surface in a difficult position for quick rescue in the churning ocean. If a diver managed to negotiate the water entry successfully and made the swim through the pounding waves to the anchor line, then the dive itself would probably not be too adversely affected; at least until it was time to decompress.

The Navy Dive Tables gave no latitude in their mandated depths and times, and the meager insurance they offered in avoiding the bends vanished once the actual dive deviated from the exact profile listed on the Tables. Trying to hang onto the anchor line at a depth of ten feet is impossible when the height of the seas matches the depth. A ten foot wave means that a diver may be momentarily pulled up and out of the water, which is completely unacceptable to say the least. The depths need to be adjusted correspondingly lower and time added to compensate, but by how much? Well, it was going to be a "wild-assed guess" anyway (to use the technical term of the day), so you might as well stay on the decompression line as long as you or your air supply could stand.

Of course, the longer one hung onto the anchor line, fighting the pull of the current and the pounding of the waves, the more tired one became. Not good, because the hardest part, by far, was yet to come; getting back aboard the boat. The *Wahoo* had a fairly large dive platform, about three feet wide, running the beam of the boat at the stern. The ladder was oversized as well, built to accommodate a diver climbing up it in full gear with his or her fins still on. But trying to climb up the ladder under the pile of dive equipment with the *Wahoo's* stern lifting twelve feet into the air, and then crashing back down - hopefully not on top of you - was next to impossible. This part could really be dangerous.

Eight to twelve foot closely spaced waves are beyond the outer reaches of any reasonably safe attempt to make a complex dive requiring decompression and lots of gear. The *Wahoo* turned back, and Craig, Gary, Hook, and I would have to wait until 1983 to touch the *Andrea Doria*. But something interesting did happen after our 1982 attempt that dramatically increased the interest in diving the *Andrea Doria* the following year. Steve Bielenda and Hank Keatts gave a wreck diving presentation at a symposium in Washington D.C. that autumn. Coincidentally, Ted Hess - the lead diver on Gimbel's expedition - was making a presentation as well, his on

SETTING THE HOOK

the *Andrea Doria*. The china recovered by the commercial venture piqued the interest of all present, but probably none as much as Steve and Hank. After the show, a cocktail party was staged for the group. Steve and Hank made a point of cornering Ted Hess to pick his brain about the location of the china, and he told them exactly where to look once inside the Foyer Deck. The *Wahoo* scheduled two expeditions to the *Andrea Doria* for the following summer in July of 1983. I signed up as crew for both.

To leave the site of the *Andrea Doria* without having dived her was bitterly disappointing. Because of the limited space due the great quantity of tanks and gear required, Steve charged most of the regular crew - except for Janet Beiser - half price for the charter. For a broke college kid like me this made it possible to participate on both trips each summer. But regardless of whether the *Wahoo* actually made it to the *Andrea Doria* or not, the $500 price for a full paying passenger was non-refundable. The *Wahoo* was unavailable for local New York wreck charters for the duration of the scheduled *Doria* attempts, regardless of whether she was hooked into the wreck or getting knocked around on a run for port; either way, the day was shot. Steve understandably had to be a businessman first, and could not absorb the loss.

Money was only the most tangible reminder of the frustration of having to turn back to port. A *Doria* expedition was not thrown together on the spur of the moment, but required planning, many extra dive tanks, and practice. To spend months - years for some - training and psyching yourself up to just to touch the famous wreck, only to be turned back so close that you could smell victory was devastating. Aborted expeditions like we experienced in 1982 were not all that unusual so far from shore at the whim of Mother Nature. But it definitely tested one's patience; it even drove some away, those not willing to risk the emotional and financial roller coaster in addition to the physical dangers and hardships.

I was fortunate. Out of my five attempts to visit the Italia liner, 1982 was our only failure, although we did have to "break" off one expedition early. In 1983, after two days and two dives on the wreck, the winds suddenly picked up to near-gale force. The waves steadily grew as we sat inside

"BLOCK-HEAD" ISLAND

the cabin of the *Wahoo* waiting out our surface interval for the second dive of the day. We waited for someone else - anyone really - to make a decision to abort their dive. We were probably all at the precipice of saying the hell with it, call it a day, and head back to port when the decision was made for us. The bucking of the *Wahoo's* bow became too much for the anchor line shackled firmly into a stout beam in the *Andrea Doria's* Promenade Deck. The line parted, and it was suddenly obvious (with a touch of relief to all) that it was time to turn toward land. Steve cranked up the diesels and we steamed for the shelter of Block Island.

Block Island, Rhode Island - if there ever was a consolation prize for an early departure from the *Andrea Doria*, it was heading to Block Island. Few returns to shore were gloomy, assuming everyone that left port returned safe and sound, but pulling into Block Island, or perhaps more appropriately in our case, "Block-head" Island, was positively festive. Block Island is the smaller, younger sister of New England's three major offshore resort centers. Overshadowed by larger siblings Martha's Vineyard and Nantucket, Block Island is just as old and historic. It is only seventeen miles from Montauk Point, and boasts beautiful beaches, panoramic scenery, and a wild nightlife. It is only six miles from the German U-boat (the *U-853*) and several other fascinating wrecks rest nearby. The USS "*Bass*," a Navy submarine scuttled in 1945 after use as target practice, and the "*Grecian*," a broken up freighter, are two other well known wrecks near Block Island.

The *Bass*, originally designated the "*V-2*," was commissioned in 1924 and had an inauspicious history. Despite four patrols in World War Two the *Bass* never engaged in combat. Ironically, a catastrophic fire in the *Bass's* aft battery room killed twenty-four of her eighty crew. She was stripped of her role as an active combatant during the war and was briefly used to haul cargo across the Atlantic. This tasking, too, was soon abandoned, and the *Bass* was decommissioned in 1945 while the war still raged on. In March of 1945, the *Bass* was sunk eight miles south of Block Island by a Navy PBY-5A torpedo/bomber using an experimental torpedo. The 342 foot *Bass* is broken in two pieces and sits on the sandy bottom in 155 feet of water. Prior to our *Andrea Doria* experiences, the *Bass* was the deepest wreck we had explored. Although the wreck of the *Bass* was a fascinating structure, the fact that it was scuttled placed it in a different category from other wrecks in my mind. The history of a vessel's sinking was an integral part of a wreck's aura. Scuttling a ship - even in a live fire exercise - tainted that history for me, lending an artificiality to the entire diving experience that was difficult to ignore. In a bizarre, perhaps unfair way, I did not consider the *Bass* and similarly scuttled vessels as being "real" wrecks.

The *Grecian* was a 290 foot long freighter resting in ninety feet of water. She had been rammed by the passenger liner *"City of Chattanooga"* in a fog to the southeast of Block Island in 1932. Her cargo of brass locks made interesting trinket scavenging, but we usually dived the wreck to supply the boat with dinner. Lobster were plentiful (at least at the beginning of the dive season), and one afternoon I speared a fifty pound cod under a broken steel plate in the crumpled folds of the bow. But by far, the most interesting catch on the *Grecian* were Anglerfish, also known as Goosefish, Monkfish, or Ocean Blowfish.

The four foot, fifty pound Anglerfish would lie flat, camouflaged on the bottom and half buried in the sand. Two-thirds of its rounded body was mouth, a mouth filled with ugly, needle teeth. It was not a good idea to spear an Anglerfish - they tended to ignore the irritating intrusion other than to turn to bite the diver holding the now empty spear gun. We had a better way to catch the fish with its delicious tail meat. After spying an Angler, the diver would empty their buoyancy compensator until they were resting heavy on knees in the sand behind the beast. With a sharp knife in each hand, the diver would take a deep breath and plunge one knife into the fish's spine and pin it to the bottom. After thrusting the second knife in above the first, the stabbing process would alternate between hands five or six times with the Angler still pinned to the bottom. After the angry critter settled down, the fifty pound fish would then be stuffed into an extra large mesh bag, a lift bag attached, and the package sent to the surface for retrieval. A lot of work (and slightly bizarre labor at that) for a fish, but Anglers were definitely good eating.

"BLOCK-HEAD" ISLAND

THE AUTHOR "CATCHING" A 50 LB. ANGLERFISH. FROM THE
AUTHOR'S COLLECTION, PHOTOGRAPH TAKEN BY HANK KEATTS (1982).

Staging a dive boat temporarily out of Block Island provided access to a wide range of fascinating and historic shipwrecks, but what I enjoyed most was returning to port after a day of diving. Block Island in the summer was the quintessential party town. Virtually every visitor from the mainland went there for one reason, to have fun, and the ferry from Port Judith, Rhode Island carried hundreds to thousands of daily fun seekers in the summer months. It was a great stopover for the multitude of sport fishing boats that would enter the harbor as well. Most would pull into the relatively small enclosure of Old Harbor, with the town's historic buildings in a half amphitheater around them, while loudly leaning on air horns and showing off their trophy catch of the day. The sport fisher's davits might be hanging a large marlin or tuna, but what got the most attention were giant sharks. The docks would crowd with onlookers as the fishermen beamed with often drunken pride.

Craig, Gary, Hook, and I always felt a touch left out. Sure, we could hang an *Andrea Doria* teacup from the *Wahoo's* davit, but that was somewhat dainty for our tastes, and there had to be a more dramatic way to create a ruckus. Unfortunately for me, my pals came up with one.

CHAPTER FIFTEEN

...but glimmering memories

November 2000, Washington State

I leaned back on my stool in the smoky, Port Angeles tavern, broke off my blank stare at the pool table and shifted my gaze to the mirror's reflection from behind the bar; with eyes dulled from fatigue and hair a tangled fray from the neoprene diving hood taken off five hours earlier, I didn't paint a pretty picture. I attempted to settle my stomach by impatiently taking small guzzles of beer. It was a recipe that worked for hangovers when younger, not exactly a hair of the dog, closer to the entire damn hide. The problem was that it was not a hangover bothering me, but instead a virus-induced exhaustion.

The day's wreck dives had been great. Chris Burgess and I pulled our way down the anchor line twice to the wreck of the *"Diamond Knot,"* a freighter lying in 140 feet of water to the west of Port Angeles on the Olympic Peninsula. The 360 foot ship had carried several million dollars worth of fish, about ten percent of the total Alaska canned salmon for the 1947 fishing season. The *Diamond Knot* collided with the coastal freighter *"Fenn Victory"* in an early morning fog and was being towed to shore when she finally sank eight hours after the accident. The wreck rested less than a half-mile from the beach on her starboard side and had been the subject of a major salvage operation soon after she sank. Broad slices had been cut out of her deck and port side to allow for thousands of the tins of canned salmon to be sucked out using a surface fed giant vacuum hose. A fairly large wreck with holds cut wide open during the salvage, the *Diamond Knot* stuck up to within seventy-five feet of the surface, shallow enough for recreational diving. The *Diamond Knot* was one of Washington's most popular dive spots.

Despite not feeling up to snuff before the dives due to a mild stomach bug, I had not dove a shipwreck in seven years and was not about to let the opportunity pass me by. Wide tidal swings made scheduling a trip to the *Diamond Knot* on a "dive-able" day a challenge, but today had been perfect. I needed to regain some wreck diving experience before my attempts on the *Andrea Doria*, but the truth was that I also needed a break and wanted to recapture the thrill and celebratory victory of a successful wreck dive, feelings now but glimmering memories to me.

Part and parcel of the way "we used to do it" was cracking a beer immediately after the last dive, well before we were even out of our dry suits. I held out longer today, but by the time the Uniflite was speeding to Port Angeles for the night the pull of old revelry returned and I joined Chris and Matt Bozak - our "boat bitch," or non-diving boat tender - in a cold one. Key to the mind-set that had kept us pursuing the challenges of the New York shipwrecks was the allure of triumphant drinking camaraderie on the return to port. It was essential to our overall state of mind, a determined toughness to persevere, whether the task before us was fighting seasickness to get into the water, or fighting a challenge on the wreck to survive.

We had not faced anything close to either problem today. Still, the reward for leaving the shore of Whidbey Island at six am, motoring three hours to the *Diamond Knot*, and then diving twice was worth a few beers. The Uniflite arced around the sandy hook that was home to the Port Angeles Coast Guard Station in relative quiet compared to the barely-contained energy of a typical *Wahoo* return to port. When the *Wahoo* would turn the corner of the Block Island Jetty at Old Harbor, the wind would be blocked creating a temporary dead calm accentuating the roar of the twin diesels under the blazing sun, which in turn would be drowned out by the scream of Wagner's "Ride of the Valkeries" blasting from the public

...BUT GLIMMERING MEMORIES

address system on the bridge. My mind wandered to our most memorable entrance to Block Island, when we beat the sport fishing boats with their davit hung sharks and marlin at their own game.

Gary Gilligan and Hook, both being in the construction trade, schemed up a plan to attract greater attention to our harbor entrance, all they needed was a volunteer. Craig was quick on his feet, and he became part of the logistics plan (he kept opening cans of beer). That left me. What the hell, why not, I thought, as Craig handed me a fresh Budweiser.

When the *Wahoo* turned the corner of Old Harbor jetty, my mind tried to make sense out of the sequence of harbor front buildings and wooden docks lined with boats. The town's layout would have clicked with instant familiarity if not for the fact that the picture was inverted. The boys had wrapped a towel around my ankles, attached a nylon web strap, and hoisted me upside down over the *Wahoo's* side by her steel davit. A bottle of ketchup was liberally dispersed across my body for added realism while the boys stood proudly at the rail brandishing spear guns over their trophy catch.

We couldn't have timed it better if we had tried. The sun-drenched spray off the water reached up into my hair while Steve brought the *Wahoo* broadside to the town for maximum effect. Just then, the ferry boat to Point Judith pulled out of its Block Island landing and quickly took a severe starboard list as nearly a hundred passengers raced to the rail to view the commotion we were creating a scant fifty yards away. The scream of "Ride of the Valkeries" couldn't drown out the frantic yells of the ferry's crew as they tried to rebalance their vessel before reaching the narrow channel exit.

We created more of a stir in Old Harbor than any shark ever did, but that was not good enough for my pals. With the *Wahoo* racing past the ferry speeding out the channel in the opposite direction, my buddies started letting out line from the davit. Oh great, I thought, as the salt spray from the *Wahoo's* wake drenched my face and made it difficult to breathe; my buddies are going to help wash off the ketchup. My head went into the water as I tried to pivot at my waist to stave off the dunking. No such luck. Three, four times I went under. I began to feel like a Salem witch, one without knowledge of what I was supposed to confess, except perhaps idiocy at leaving myself at the mercy of my friends. Fortunately, the commercial dock was close and my pals eventually hauled me in as the fenders went over the side.

Our grand entrance to Block Island signaled the start of an afternoon and evening of festive mischief. First stop was always "Ballards," a dockside ballroom proportioned restaurant that entertained hundreds of patrons with its island bar and joke-cracking singer "Moriarity" and his Irish band. From a drinking and singing perspective, Ballards was undoubtedly the premier fixture on Block Island.

In 1986, a year after leaving New York for the Navy, I received a small package in the mail from my east coast buddies. Reading the note in horror, I learned that Ballards had burned down. No one had been hurt, but the entire bar was nothing more than charred rubble. The box contained a smoke-blackened outdoor water spigot, a final artifact recovered from the "wreck" of Ballards. I held on to the spigot with the fear of last call in my belly. Thankfully, Ballards was rebuilt the following year.

After a warm up at Ballards, we would dare out on forays into town and the rustic island countryside. We were not exactly well behaved, but the authorities were great sports about not confusing loud mouthed, good-natured obnoxiousness with a real threat. We never started fights or broke anything that we did not pay for. Well, not usually at any rate.

Our group scoured the countryside on rented mopeds one sunny afternoon, occasionally straying from the road and traversing open fields. Gary was the last in trail, a particularly vulnerable position, one which led him to become a sacrificial lamb after one particularly egregious run onto private property. A police car appeared from nowhere with lights flashing and a not-so-blazing high speed chase ensued. Gary whipped the throttle grip of the single horsepower moped wide open as he tried to make a run for it. We watched from a distance, saw that his pathetic attempt at fleeing the scene of the crime was futile, puffed out our chests, and did the manly thing - we deserted our buddy and got the hell out of there. The last thing

...BUT GLIMMERING MEMORIES

we wanted was a jail bound disruption to our well thought out day of bar hopping.

Finally realizing that the jig was up as the cruiser slowed down beside him, Gary calmly followed the uniformed officer's instructions to get into the back of the patrol car, while a civilian clad gentleman sat silently in the cruiser's front passenger seat. Once the doors were closed the interrogation began. The uniformed driver politely delved into Gary's psyche for the root causes, the fundamental societal failings that had steered the young lad wrong. Gary was eager to confess, to spill his guts (he had been caught red handed after all), but the civilian clothed gent in the front kept interrupting his answers with brusque questions of his own. Gary finally grew impatient with the talkative passenger and instructed him to please "shut the fuck up," so that he could answer the officer's questions. The patrol car became deathly silent, the conversation not resuming until five minutes later when the man in the front seat was dropped off at a lone country dwelling. Next to the mailbox was a modest sign that read "Chief of Police." Oops.

Gary was taken downtown where the official charges, verdict, and punishment were read in one quick sentence: reckless operation of a moped, trespassing, resisting arrest (an embarrassing interpretation of Gary's attempt to flee), and lastly, insulting the Chief of Police - guilty on all charges, fifty dollar fine. The only problem was that Gary's wallet was empty and all he could come up with was the promise of the fifteen dollar moped rental security deposit assuming that he could get a ride to the crime scene to retrieve the thing. After several minutes of skillful negotiation, Gary bargained his way down to a fifteen dollar fine and a van ride back to the moped. He joined us in Ballards barely an hour after being arrested. The cops on Block Island were excellent sports. They simply wanted to ensure that no one got hurt and property damage was kept to a minimum.

Sitting in Port Angeles, with my stomach rumbling and my head still relatively clear, I felt a mere shadow of my former self. It was only eight o'clock, but I called it a night and walked back to the boat hoping to recoup my wreck diving prowess more quickly than that of the post dive partying.

SETTING THE HOOK

My training was making progress, but it seemed to be going too slowly all the same. I had completed a "nitrox" course at the local dive shop in Anacortes and used the new gas mixture during the two *Diamond Knot* dives. Nitrox took advantage of the relatively simple concept of reducing the decompression penalty on a dive by replacing nitrogen in the breathing gas with oxygen. Instead of breathing air, with its 21% oxygen content, our *Diamond Knot* dives used a mixture with 28% oxygen. The appropriate oxygen content for the dive would be programmed into my new wrist strapped nitrox computer, and the enriched gas mixture allowed for a longer and deeper stay on the wreck than would otherwise have been prudent if breathing air.

The drawback to nitrox was the limitation imposed by the increased oxygen content. Breathing nitrox did not feel any different from breathing air - there was no taste, no discernable change in the density of the gas, nothing besides the label on the side of the scuba tank to indicate that it held something other than air. The higher proportion of oxygen in the gas blend resulted in a higher oxygen partial pressures at depth, and breathing nitrox could kill at depths where it was still safe to breathe air. While the safe oxygen partial pressure limit for air occurs at about 220 feet deep, increasing the oxygen content to 28% reduces that safe depth to roughly 155 feet; quite a difference. If a scuba tank filled with nitrox was confused with another filled with air, a diver could experience the uncontrollable underwater convulsions of oxygen toxicity and risk drowning.

It was therefore essential to accurately analyze the contents of every tank before a dive. The filling facility would use their on-site oxygen analyzer to positively verify gas content before returning a tank to the customer. This calculator sized contraption connected to the tank and directed the flow of gas through a portal and an oxygen sensor, which translated the derived content into a numerical percentage. The customer would duplicate the process again as a cross-check when the cylinder was picked up at the dive store. The specific percentage of oxygen in a tank and corresponding maximum safe operating depth would be written directly onto a label attached to the cylinder to avoid any possible confusion later.

...BUT GLIMMERING MEMORIES

Diving with nitrox was a poignant example of how a little common sense and attention to detail could avert potential disaster. Planning for a nitrox dive was simple, but critical. As sport diving became more technically advanced and complicated over the years, the need for a disciplined attention to detail increased exponentially. There had been and always would be "gotchas" underwater, but for each advance designed to reduce the effects of one "gotcha," it seemed that several others were introduced.

The *Diamond Knot* dives were fun, but nothing particularly challenging. We grapneled into the bow section of the freighter forward of the broken up cargo hold at 125 feet deep, and if not for the lingcod and wolf eels, it would have been almost barren. We dropped down the anchor line together, but Chris did not have double tanks or a desire to decompress and he left the wreck to ascend much earlier than me. I spent most of each dive close to the anchor line, trying not to lose sight of it for more than a few minutes. The current was strong and growing, and I did not want to be forced to decompress away from the anchor line and at risk of being swept out to sea during a free-hang decompression.

Virtually all shipwrecks are disorienting to some degree. Better visibility does much to alleviate this confusion, but the passage of time and deterioration of steel bulkheads and decking can eliminate most recognizable features after a ship has been on the bottom for several years. There are a few tools at a diver's disposal to find the way back to the anchor line. The depth where the hook is set is a key factor because it is usually the only quantifiable element of navigation available. If properly assessed and remembered, the depth at the tie-in can eliminate doubt as to the line's location in the up and down, vertical dimension of the wreck. A compass might help, but the surrounding metal could easily cause erroneous readings. A diver needed to take a calculated and measured observation of the wreck's features. Which way did the structure seem to run, were there any obvious "landmarks" nearby, a crane, rail, or a significant change in depth? For each landmark noted, there were usually other, similar features to confuse a diver in the anemone festooned, rust encrusted hulk. It took practice and discipline to maintain a sense of location on a shipwreck, and I was out of practice.

SETTING THE HOOK

I freed the hook at the end of the dive and wrapped the chain around the grapnel so it would hang upside down, making it less likely to re-catch on a piece of wreckage. The current eased momentarily and I inflated my buoyancy compensator, hoisted up the anchor grapnel and dropped it on the highest, flattest piece of wreckage I could find before heading up the anchor line. The current started moving quickly again after reaching the twenty foot decompression stop, and with relief I felt the hook bounce free under the tension of the anchor line and drift clear of the *Diamond Knot*. Overall, they were fairly good practice dives, but not as challenging or rewarding as my last time on the wreck seven years earlier.

In 1993, the charter boat had hooked into the *Diamond Knot's* superstructure. I found an opening in the wreck, tied off a penetration line to make up for my unfamiliarity and lack of a buddy, and entered the wreck. After swimming fifty feet into the disheveled interior, I came across a thickly glassed, elongated eight-inch light housed in brass. Using a pry bar, I worked the fixed light loose from the bulkhead and stirred up a quite a bit of silt in the process. Fortunately, the brass light was high enough off the bottom that a few feet of visibility still existed in the upper portions of the chamber. After hooking off the penetration reel temporarily to my weight belt, I used both hands to pull the light free, cut the attached electrical wires loose with a dive knife, and put the artifact in my catch bag.

Continuing further into the wreck, I swam until catching the gleam of white reflecting off my dive light. It turned out to be simple white china, and I retrieved eight pieces of the rugged dinnerware from the silt by feel before reeling in the penetration line for my return swim through the zero visibility blackness. The dive reminded me of the New York wrecks, at least how they had been eight years earlier, and in a strange way it felt a bit like home. On the boat deck I took out my prizes and looked them over. They were nothing close to the ornate beauty of the colorful *Doria* china, but still, it had been my first dive inside the *Diamond Knot*, and I had gotten a bag full of artifacts.

...BUT GLIMMERING MEMORIES

I had not felt that invigorated underwater since my last dive on the *Andrea Doria* in 1984. Hook and I had dropped feet first through Gimbel's Hole into the blackness of the Foyer Deck that morning, going down and down until the tiny luminescent hand on my depth gauge read 205 feet. Feeling the darkness creeping at our shoulders, threatening to come round full circle to snuff out our feeble dive lights, we slowly, with exaggerated caution in our every move, finned and pulled our way aft, staying above the thick sediment and moving around the collapsed debris impossible to identify. We went past the mesh of threatening electrical wires, the maze of jumbled beams tossed like a child's pickup sticks, past the bathtub sitting sideways in the narrow passageway, and over the remnants of one more bulkhead to just beyond where the "floor" dropped away into a sideways stairwell in the wreck's new orientation. After a brief confirming glance at each other and our surroundings, we prepared mentally for the impending oblivion of silt, referenced the "wall" to our right, which was really the ceiling, let the air out of our buoyancy compensators and dropped to our knees.

Ten minutes later we were rising up in the relative open nothingness of the Foyer Deck, still inside the *Andrea Doria*, but feeling like we had just been set free from a Helen Keller-like sensory deprivation impossible to truly describe. We floated up, keeping close to the side of the corridor so we could exchange glances from depth gauges to the reference of the bulkhead, but not so close to risk becoming caught by the wreck, cheated just as we were about to make a clean get away.

An amazingly bright world bathed in sunlight welcomed us as we emerged from the heart of the *Andrea Doria* and swam to the anchor line. We had a combined total of fourteen pieces of the china in our mesh bags, and the current swept away remnants of clinging mud from the plates while we hung onto the anchor line and decompressed. I kept looking down as gradually more of each trophy was revealed, the gold of the inlay, the crown, the reddish painted braid on the rim, the insignia "Italia." I checked again and again that the bag's clip was secure at my waist, that it would not release and plunge out of sight downward forever.

SETTING THE HOOK

We were done for the season on the *Andrea Doria*, but we would return the next year, we promised each other, to do it again, to feel the thrill and victory of the ultimate dive. Why wouldn't we?

CHINA FROM THE *ANDREA DORIA*'S SECOND CLASS DINING ROOM. ALTHOUGH SIMILAR IN APPEARANCE TO THE ORIGINAL "DISH HOLE," ACCESS TO THIS PART OF THE WRECK WAS NOT AVAILABLE UNTIL SEVERAL YEARS AFTER THE AUTHOR LEFT NEW YORK. PHOTO COURTESY OF BRADLEY SHEARD (1991).

CHAPTER SIXTEEN

...jolted by the shock of her life

August 1985, U.S. Naval Aviation Officer Candidate School, Pensacola, Florida

"On your face, dammit. Get'em 'til I'm tired a watch'n you!" barked the gravelly voice from beneath the wide brimmed Smokey the Bear hat.

Twenty young men, blank-faced in a futile attempt at anonymity, with heels stuck together and backs rigidly straight, shouted out the pushup count in unison. A mud puddle from a grass sprinkler slowly evaporated in the one hundred degree sun to the left, but our Drill Instructor appeared to be in a good mood and we were allowed to press on with our afternoon punishment in the dried out dirt.

For the past eight weeks our class of forty-three fledgling Navy Pilot candidates had been under the gentle tutelage of Staff Sergeant Gerhardt, United States Marine Corp. The class had started together mid-June and now, with only six weeks left, less than half the original faces remained. Interspersed in the regime of routine physical abuse were survival training, marching and close order drill, and classes in aerodynamics, navigation, and jet engine mechanics. Several hours of advanced brass polishing rounded it all off. Whatever it took to make certain one was miserably fatigued for every decision contemplated.

SETTING THE HOOK

The goal was to inculcate the discipline, attention to detail, and respect for authority that we would need to survive flying U.S. Navy jets by tearing us down and then building us back up in a proper military fashion. The mantra, repeated endlessly, was "attention to detail is the key to survival; discipline is the key to success." Aviation Officer Candidate School would transform us lowly, civilian brain-dead slugs into Naval Officers in just fourteen weeks. This seminar in extreme discipline and skivvy folding was a prerequisite to the challenge of two years of flight school that would in turn open the door to frequent and long aircraft carrier deployments. Here I was, strictly my choice, able to quit and leave at any time; all those years of college and still an idiot, my New York dive buddies must be chuckling.

Our class had "secured" in the program's sixth week and we could look forward to Saturday nights in town, but only while in uniform and under casual shore patrol surveillance. Our only other glimpse of the world outside the Pensacola Naval Air Station would come at mail call. Once back in formation, drenched in sweat and carefully poised with proper military bearing, the mail was distributed with elaborate mock-ceremony. Staff Sergeant Gerhardt, United States Marine Corp (the name given him by his mother, we were certain), would growl out a name, the lucky candidate would march up, stand at attention, and wait for the letter and accompanying punishment certain to be doled out for screwing something up along the way.

Finally, it was my turn, and after a rigid march and square up in front of the drill instructor, I managed to about face with the letter in my teeth after a mere fifty pushups; Staff Sergeant Gerhardt, United States Marine Corp, was indeed in a good mood. We were left alone for a few brief minutes to read our correspondence.

The letter was from Hank Keatts, my dive buddy professor. He had taken time out over the course of several weeks of diving to jot down events and fill me in. I started to read:

...JOLTED BY THE SHOCK OF HER LIFE

Pete,

21 July - 9am - we left the dock at 11 pm (two nights ago) and tried to get to the Doria. However, in heavy seas, the starboard engine exhaust sprung a leak and we were forced to go into Block Island. We arrived at 2 am and Ballard's was closed, so we crashed. It was determined the following morning that the exhaust could not be repaired. In an emergency the engine could be used but the exhaust fumes would enter the aft cabin. Steve decided to run on one engine and we left Old Harbor at 8 am and arrived over the wreck at 5 pm. I made a dive at 6:30 with Gary Gilligan. Don and Craig fished (they caught several Pollock). Gary and I went along the Promenade Deck looking for windows, but did not find one. It was rather dark. We then went to Gimbel's Hole and pulled the grapple hook out (it was chafing) and left it on the hull. The grapple had been thrown right into the hole. Gary Gentile and John Lachenmeyer had taken a line down and fastened it to the Promenade Deck as usual. They then went into the Dish Hole and picked up several dishes.

Most of the people are now suiting up for their morning dive, but I am going to wait and dive around noon. Five of us hanging on the anchor line yesterday was a bitch. Gary and I had to hang at 40, 30, and 20 feet, there was no room at the correct depths. Also, I would rather let the sediment settle in the corridor before I enter.

3 pm – Craig, Don, and Sally went to the Promenade Deck and came up empty. John Lachenmeyer and Gary Gentile brought up a few plates and cups and some of the others found one or two. I went in alone at 1 pm. Everybody else had been down. The other divers had been complaining about the anchor line's scope. Some of the guys said it took them 6 minutes to reach the hull. They were right. I made it down in 5 minutes, but I was so tired from pulling myself (a current was running) down that I felt as if I should return to the boat. If I were intelligent I would have. Anyway, I dropped through Gimbel's Hole as I have done several times before, but this time I was disoriented as to where the corridor was. While I was trying to orient myself I realized that I had become entangled with monofilament fishing line. I went back up to the hull and removed my knife. It was necessary to cut the line in at least 3 places before I could free myself. Then I dropped back into the hole and immediately found the corridor. There was little bottom time left so I picked up a plate and a very large silver serving tray. However, a dive that starts as screwed up as this one usually continues as such. When I reached my first stop I dropped the tray. The 40 minute hang was as much fun as you remember. I was using Steve's buoyancy vest and tank bands, so the straps pulled as

tight as I could get them were still loose. The tanks were working up over my head with each surge (of current and waves). My dry suit hood was frayed at the seam due to the constant friction. The only consolation was that I was not attempting to hang with 4 other divers (like yesterday). The wreck is much more spooky when you are diving alone.

So, after 2 dives I have a grand total of one dish (the Italia insignia can hardly be read). I am sitting here wondering why I even came. It is easily understood why I do this only every other year. If the 250 feet of scope is left out for tomorrow's dive I will pop a Miller Lite and sit in the bleachers.

7:30 pm — The sea has really laid down. What a difference it makes. After a screwed up dive and six pack of Miller Lite things do not seem so bad. I hope to sack down soon, so that I can make my 3:30 am — 4:45 am radar watch (for ships). I know that you wish you could cover my watch for me.

22 July 9:00 am — After the radar watch I stayed awake to get ready for my dive. I wanted to be the first in the water. However, my dive partner was slow getting into his gear and 2 other divers entered first. I suggested we wait a few minutes to let the sediment clear (they were going after dishes too). At 6 am we were ready to slip on our tanks when a storm blew up and Steve canceled all diving. When the first two divers finished their hang they kissed the boat deck. The current was so bad they had to hold on with both hands and had trouble monitoring their gauges. Thank God I didn't jump in before Steve stopped us. I was lifting my tanks onto the doghouse to put them on.

What a horrible weekend. Be glad you were not here. Don and Craig made only one dive and that was to the Promenade Deck. They have no artifacts to show for their effort. I retrieved only the one plate. Maybe you had better come up and re-educate us. I am not looking forward to next weekend's trip. I think I will try to go back to sleep and miss some of the 10 hour boat ride.

3 pm — The autopilot is not working due to the heavy seas and Gary Gentile woke me up to steer the boat for the next hour and a half. You know how I love boats and all things nautical. We are having some fun now.

30 July — We dove the Bass last Wednesday. However, we were on the bow section instead of the stern. Nevertheless it was a good dive. As usual I drank too much that night at Block, but I did not feel too bad the next morning. We dove the U-boat the following morning and I found a one-man life raft inside. The raft inflated after 40 years under water.

I was too tired to go on the second Doria *trip. Dillon (last night) said Steve and Gary Gentile went to the Gift shop and found a large amount of different artifacts. Gilligan followed on a later dive and also found some nice artifacts. As usual I made the wrong decision. I should have been there.*

Do not bother to answer this letter. I will be spending the month of August in Texas and would not receive your reply until September.

Good luck, and keep up the good work,
Hank

I stuffed the letter back into the envelope and took a minute to think about the world I had left behind. By the time I was able to phone Don, the Hook Nosed Bastard, the following Saturday, there was no shortage of news to tell. Another diver, this time a guy from Florida, had died on the *Andrea Doria* and had almost killed Sally Wahrmann in the process. Hank had strategically omitted the fact in his letter; maybe he figured I had enough on my mind. Sally's intelligence was only rivaled by her calculated calm. She would dive the *Doria* with double seventy-two cubic foot tanks of air, stay underwater for the same duration and depth as the guys, and still come up with half her air supply. Steve would joke that she was the only person he knew to dive the *Doria* on a single seventy-two, not a particularly large style of tank, and a ridiculously insignificant amount of air to breathe for twenty to twenty-five minutes at over 200 feet deep and still have sufficient air remaining for the extended decompression. The amount of air required on a dive could vary tremendously depending on the individual diver's comfort level and technique. Sally was cool and relaxed, and that was what kept her alive.

Sally had been on a dive to the Gift Shop with Gary Gilligan when the event occurred. The entrance to the Gift Shop turned out to be directly below the Dish Hole at 220 feet along the very edge of the Foyer Deck shaft, which was information that would have been nice to know the previous summer when I was literally wasting my breath looking for it in the wrong direction. Sally had given Gary Gilligan several minutes head start to get situated before she joined him digging for artifacts. When she reached the entrance to the Gift Shop, she hung neutrally buoyant in the Foyer Deck shaft clear of Gary's billowing silt and looked for a good spot to dig. She had not joined him for more than a minute when she was jolted by the shock of her life.

Something hit Sally from directly above with enough force to nearly knock her senseless. She suddenly found herself tumbling uncontrollably into the black pit of the Foyer Deck, accelerating in her drop as she grew more negatively buoyant under the increasing pressure. The impact had knocked her regulator out of her mouth, twisted her dive mask half off her head, and flooded her face with icy water. Fortunately, she had taken Steve's advice from years before and worn her mask strap beneath her neoprene rubber hood. If not for that precaution, the dive mask would have been wrenched completely off and probably lost forever. Without a dive mask to see, Sally's chances of survival would have been negligible.

Sally's out of control downward crash was only stopped when she hit the rubble pile flat on her back at the bottom of the Foyer Deck shaft. Although the sandy bottom surrounding the *Andrea Doria* is 240 feet deep, the digging action of the ship's hull in the sand had created a wash out directly beneath the wreck. Sally attempted to regain control, starving for air with her regulators floating somewhere nearby, blind without a mask, and severely narced at 252 feet deep far inside the *Andrea Doria*. Most divers - even of the caliber to dive the *Andrea Doria* - would almost certainly have drowned. Sally fought to maintain control, reached out with searching hands, and finally found one of her regulators. If not for her discipline and experience, she would probably have drowned in a matter of seconds between the narcosis and the urge to hyperventilate. Sally reset her mask, cleared the water, and took stock of the situation.

Later, Sally would explain the experience in this way: "I really was knocked silly. I lost my regulator and my mask was hit sideways. By the time I got control of myself, I was sitting on the bottom inside the wreck talking real loud to myself - 'you're okay, whatever hit me is gone; fix your mask; you have plenty of air; which way is up? Calm down!'"

Controlling one's basic physiology under these conditions is an incredibly difficult task. Sally's body was telling her to breathe quickly while her instincts screamed at her to escape the darkness, to start swimming anywhere to find an exit and shoot straight to the surface. Regardless of how many classes a diver might attend or how many books they might read, there is only one factor that can prepare them for such a circumstance: experience. Fortunately, Sally Wahrmann was immensely experienced. She stopped moving completely, fought the nearly irresistible urge to bolt, and simply breathed. Slowly, deeply, and with concentrated effort, Sally regained control of her breathing. She looked up. Eighty feet above her, almost invisible through the cascading silt from Gary Gilligan's digging at the gift shop, she could barely make out a small patch of light the size of a quarter, the sole visual clue that there really was an exit from Gimbel's Hole. With focused deliberation she inflated her buoyancy compensator and began a controlled ascent.

Sally had not planned for a dive to 252 feet and she extended her decompression to two hours to compensate, an option only made possible by her impressively low breathing rate. During her long hang, the question plagued her; what hit her with such force? The disturbing answer was that it had been another diver. The diver who plunged out of control into Sally was a first timer to the *Andrea Doria*, but not a novice to wrecks or great depth. He had over fifty dives on wrecks of comparable depth, and was capable of free diving on one breath of air to an incredible 170 feet. But what he set out to accomplish that July morning was a far different challenge than those he had achieved in the warm, clear Florida waters.

The *Wahoo* charter had been organized by Captain Spencer Slate of Key Largo's Atlantis Dive Center. Ironically, Captain Slate ended up not making the 1985 charter after getting bent on a warm-up dive on the wreck of the "*Wilkes Barre*" off Key West. Several of the divers who did go on the charter were well known experts: Rick Frehsee, world renowned underwater photographer; Neal Watson, Bimini resort owner and the holder at the time of the world depth record while breathing air (an incredible 437 feet); and Billy Deans, the future father of the open water trimix technical training curriculum. Accompanying them was a cadre of primarily tropical water divers with a tremendous amount of deep-water experience, but as they were to learn, the cold and dark *Andrea Doria* was a world apart from warm water wrecks.

Sally Wahrmann was acting as Steve's appointed Dive Master for the expedition. Her responsibilities were to verify appropriate certifications, review passenger logbooks for requisite experience levels, and to conduct the pre-dive safety briefings. It was her job to accomplish a hands-on verification of the passenger's required experience as it had been related to Steve over the phone prior to being permitted on the charter. If anything was suspect, the offending passenger would not be allowed to dive. Once in the water, however, the divers were on their own.

Not everyone was happy to have their logbooks reviewed, some divers viewed it as an insult to their pride, and this was particularly true with John Ormsby, the unfortunate diver who was to plunge into Sally and almost kill her. Sally put on her CPA hat (Certified Pain in the Ass), and with Steve's firm backing stood her ground and demanded to see Ormsby's logbook. Finally, he relinquished the record, but only after it was made abundantly clear to him that it was the only way that he would be permitted to dive. Confidence is importance on a dive like the *Andrea Doria*, but arrogance can be fatal.

The full story slowly unraveled. Strange things had happened starting with the very first dive of the expedition to set the hook. After arriving at the dive site, Craig and Hook followed a light-weight polypropylene rope left tied to the wreck on the previous expedition and shackled the *Wahoo* into the Promenade Deck with a heavier line. The two then dropped down through Gimbel's Hole to the Gift Shop, kneeled on the small ledge and quickly stirred up the silt to nearly zero visibility. After twenty minutes on the wreck, Craig stuck his light in Hook's face, signaled that it was time to go, and then began to slowly kick his way up the shaft in an effort to save air by not immediately filling his buoyancy compensator. As Hook turned to follow, one of Craig's kicks smacked him in the side of the head, tearing the mask from his face. With his face shocked by the icy water, Hook rested his knees precariously on the Gift Shop ledge, struggled to regain control of his breathing, and blindly reached out to find his mask. Miraculously, his fingers found the slowly sinking mask before it dropped to the bottom of the Foyer Deck. He regained his composure, put the strap back around his hood, and cleared the water from mask. From that day forward, Hook wore his mask strap beneath his hood. Hook rejoined Craig and they started up the anchor line.

Rick Jaszyn planned to follow Craig and Hook after they had tied into the wreck and then penetrate solo into the Dish Hole in search of china. Sally Wahrmann and Gary Gilligan were also close behind, but they had opted to go to the Gift Shop. To have more than one team of divers at a time in the confined space of the Dish Hole was difficult at best, and surface-made plans incorporated the time needed to let Craig and Hook exit the Dish Hole and the silt to clear before Rick entered.

Gary planned to gently dig at the Gift shop just beneath a small overhang, while Sally did the same at the very edge of the space no more than five feet away. This put Sally in a position where she would need to be neutrally buoyant for a time, hanging midway down the Foyer Deck shaft suspended thirty feet above the collapsed pile of rubble at the bottom.

Far above them on the deck of the *Wahoo*, the frenetic activity of divers gearing up reached a near alarming pace, and Billy Deans, John Ormsby, and Lou Delotto were the next to clamber over the side. Billy Deans had convinced Steve to allow him to hang a long oxygen hose from a cylinder at the bow of the *Wahoo* leading to a fixed decompression station. This was the first use of oxygen for accelerated decompression on the *Wahoo*; perhaps the first on a northeast scheduled charter. The team planned on rigging the underwater decompression station at the beginning of their dive, and only then would they continue down to the *Doria*.

John Ormsby entered the water and in a disturbingly prophetic event quickly became entangled in the rigging lines of the "deco" station. Billy Deans signaled for Lou Delotto to start down the anchor line, while Deans assisted Ormsby in freeing himself before the pair made their descent and caught up. The three planned a conservative dive to survey the Promenade Deck and acclimate to the new environment before venturing inside the wreck on a future dive, but John Ormsby had other ideas. As soon as Ormsby was free of the deco-station, he quickly swam down the *Wahoo's* anchor line, so quickly in the absence of current that he passed behind Lou Delotto, probably halfway down the anchor line, without being seen. He headed directly for the rectangular cut-out of Gimbel's Hole.

A reasonable hypothesis of what transpired next was that Ormsby, probably over weighted in the unfamiliar confines of a dry suit, ignored the dive plan made with his buddies and instead swam directly to Gimbel's Hole in search of china. Leaving his two buddies far behind, he pushed out over the edge of the cut-out to Gimbel's Hole, shot downward head first and out of control into the blackness, and accelerated as any remaining positive buoyancy in his dry suit was squeezed to virtually no effect.

After the *Andrea Doria* sank, one of the first things to deteriorate was the wood and thin metal bulkheads that housed the ship's wiring. The loose debris dropped to the bottom, but the cables exploded in their new found freedom with ends still securely anchored to their bulkhead runs. In his out of control descent, Ormsby might have grabbed at the cables that surrounded the mouth of the Dish Hole in a desperate attempt to stop his downward plunge, or he might have inadvertently flailed directly into their web. Regardless, the cables had been a serious concern for years and knowledgeable *Doria* divers gave them a wide berth when entering the Dish Hole.

John Ormsby buffered his fall by ramming full speed into Sally Wahrmann and only came to a complete halt as the wires stretched taut and then lifted him back up like a spring. By the time Sally found her regulator, cleared her mask, and exited the *Doria*, the Foyer was so stirred up with silt that neither she nor Gary Gilligan ever saw Ormsby entangled in the cables. Neither did Rick Jaszyn when he exited the Dish Hole at approximately the same time. When found, John Ormsby was hanging upside down just below the Dish Hole hopelessly wrapped in a net of electrical wires.

There was another problem. John Ormsby had attached to his weight belt an array of open-faced brass clips to hold tools. They were known in the northeast wreck diving community as "suicide clips." For years, Steve had vigorously tried to dissuade divers from using the clips. The clips required only a slight pressure against their face to open and capture lines, or in this case, wires. I used to utilize one of the clips myself - they were easier to manipulate with heavy gloves than the more secure safety boat clips. The single one that I had used in the past was for my extra large mesh bag, however, which was high on my waist in front of me where anything caught in it could easily be accessed. John Ormsby's suicide clips ran around his weight belt and out of easy reach behind him.

Billy Deans went looking for Ormsby after losing sight of him in his race to the wreck and eventually found his buddy wrapped tightly in the web of wire. It was only later that the wires caught in Ormsby's belt suicide clip snaps would become visible. Unable to extricate Ormsby, who was unresponsive to Deans' attempts to give him air, Billy Deans ascended up the anchor line directly to the surface without decompression in an act of heroic desperation to signal to the *Wahoo* crew that his buddy was trapped in the Dish Hole. Perilously low on air, he went back down again to try and help Ormsby.

When a frustrated Billy Deans ascended the anchor line for a second time he handed a slate to Sally and Gary while they were finishing their thirty foot decompression stop. The slate read, "Help Buddy Caught in Wreck." Sally and Gary motioned to Craig and Hook at their ten foot decompression stop above them. Craig Steinmetz descended to read the

SETTING THE HOOK

slate, broke decompression, surfaced and swam to the boat to relay the urgent message, unaware that Deans had already alerted those onboard the *Wahoo*. Craig went back down to finish his hang with the nagging anxiety that his interrupted decompression might cause the bends.

The crew on the *Wahoo* raced into action. In scant minutes with Janet's vigorous assistance, Gary Gentile was suited up and dropping over the side for the race down to the blackness. He brought with him a spare tank and air, but it was far too late - when Gary Gentile reached Ormsby, he found him dead with both tanks completely empty. It took Gary Gentile and Rick Jaszyn, both *Wahoo* crew on the expedition, multiple dives utilizing bolt cutters in the blackness of the Foyer Deck, surrounded by the same wires that had proven to be so deadly, before Ormsby's body could be freed. Finally, Gary Gentile clipped a lift bag to Ormsby's stiff form and sent the body to the surface. But the ordeal was not yet over.

When the tethered lift bag reached the surface, Craig and Hook sped out in the chase boat to retrieve the body. It seemed Craig and Don were making a morbid cottage industry out of retrieving corpses on the surface, this being the second year in a row following Kennedy's death in 1984. Craig grabbed the lift bag while Hook took hold of Ormsby's body. As they were lifting, Ormsby's arm, now stiffened by rigor mortis, swung upward and smacked Craig sharply on the side of the head, causing him to yank sharply on the lift bag as he fell backward into the inflatable. The lift bag was too light for the additional strain and its snap broke. The gruesome package dropped back down to the wreck. Fortunately, Steve had asked that a line be attached to Ormsby's body and secured to the wreck before being sent to the surface. As it turned out, Ormsby floated down directly onto the hull of the *Andrea Doria*. Several of his friends went down to the wreck, relocated the corpse, and sent it to the surface once again.

Just days before the John Ormsby tragedy, Chris Dillon had made a dive to the same spot. I read the chilling excerpt from his logbook sixteen years later.

"*Gary Gilligan and I pass Steve Bielenda and Gary Gentile as they are going up the anchor line. Gary and I hit the Doria in two minutes. I unfold my catch bag with Gary to my right next to the red dotted visual line we dropped into the Dish Hole (to follow down — a new safety precaution in 1985). At 210 feet on my depth gauge with my light and catch bag in my left hand I am looking for a landmark but see nothing except a strand of three cables and an "I" beam running down the hole. I reached out to stabilize myself by grabbing the "I" beam as I am still dropping into the hole. The "I" beam tears loose and almost falls on me; it scared the hell out of me. As I pushed it away, I grabbed the strand of three cables to stabilize on them. Christ — they are loose and fall away. I am still dropping — 215 feet. I now shoot to the wall and hang on for my life. I am absolutely terrified. To quiet myself down, I kept my eyes closed and concentrated on my breathing — slow down — get control — OK — it's coming around — keep still and breathe easy. OK — control is back. Now my breathing is regular and easy, but my heart rate is still pounding and my head is not ready to process any new activities. All I'm doing is holding on to the wall. It seems like an eternity, I can see Gilligan's light flashing as he works the Gift Shop shelves about eight to ten feet to my right. As all my senses returned to normal, I checked my air gauge: 800 pounds in one tank, 3,000 pounds in the other, only twelve minutes into the dive. I can't get to the Gift Shop now, it's too late to go over there and try to grab up anything. Screw this. All I want is to get the hell out of here. I tap Gary and signal up. I realize now that when I let go of the wall I dropped my goodie bag. Ah, well — the* Andrea Doria *gives a little and takes a little."*

What Chris Dillon gave away to those three strands of wire on the *Doria* that summer was insignificant, what he gained was inestimably valuable - experience. John Ormsby never had the opportunity to return from the *Doria* with any treasure of material or knowledge, and what he left was infinitely precious.

In 1981, Steve's primary competitor, the charter boat *Sea Hunter*, had a diving fatality on the *Doria*. Little was known about the circumstances that led to this particular tragedy other than the fact that the diver was found inside the wreck, not entangled in any way, and with air still remaining in his tanks. He drowned, probably due to deep-water blackout, but it was mostly conjecture as to exactly how or why it happened. The *Wahoo's* 1985 expedition brought the number of *Andrea Doria* deaths to three since Gimbel's 1981 project opened her up literally and figuratively to regular visits from sport divers.

SETTING THE HOOK

The *Wahoo's* second expedition to the *Andrea Doria* marked the end of her trips to the Italia liner for that year. I finished my quality time with Staff Sergeant Gerhardt, United States Marine Corp, feeling that perhaps I was in more comfortable surroundings in boot camp than my New York friends had been over the *Doria* during that summer in 1985.

CHAPTER SEVENTEEN

The New Way

April 2001, Washington State

Ron Akeson hesitated in a hover five feet above the sandy bottom at the third dock pylon and waited for me to catch up. I saw that there were no bubbles coming from the two regulators or valves on the twin tanks on his back, and watched him slowly turn to allow me to complete the inspection with a look at his two "stage" bottles clipped one to a side. John Campbell was looking over my gear in a similar manner. Once complete, John swam past both of us and temporarily took the lead while Ron gave his equipment a quick once over and then finned back to the front. It took all of ten seconds before we were continuing our descent along the gradually sloping bottom out into Possession Sound.

The floating dock normally attached to the heavy duty metal pylons had been removed for the winter to avoid storm damage. The Mukilteo ferry's rumbling engines and churning propellers were distinctly audible underwater as the ship left the pier three hundred yards to the north; the car ferry made near constant runs between the mainland and Whidbey Island's southern tip, only thirty minutes away. The State Park adjacent the Mukilteo ferry terminal was a popular deep diving training spot.

Ron veered sharply to the right at twenty-five feet deep and looked for the thin nylon line anchored in the sand by the foot-long metal corkscrew that Ron and John had set three months earlier. The bottom was of such a slight gradient that sighting a compass course alone could waste valuable bottom time as a diver's swim wandered while trying to distinguish the contour - the line was a vital visual reference in the midst of a desert nothingness. My light focused on the thin nylon rope, about the diameter of a string, and followed Ron's fins while John's light reflected off the bottom in my peripheral vision.

It had only been on the last two trimix training dives that I began to feel truly relaxed with the nearly 250 pounds of gear strapped behind, around, and underneath me, but it was not until exceeding my long time depth limit of 220 feet that my confidence level was completely restored. On the last training dive we had descended along the same guideline to 240 feet, struck out perpendicular to the line for fifteen minutes, and then returned for a lengthy ascent and decompression, spending a total of one hour and forty-one minutes submerged in the forty-two degree water. Ron signaled at the end of the required decompression to deploy the lift bag attached to the end of my penetration reel. My fingers were so numb from the cold that it took an extra sixty seconds to simply turn the friction knob on the reel to release the tension on the thin line.

This was my graduation dive, the last in a string of ten underwater sorties to update my training and practice the new aspects of technical diving. Four of the dives had been staged from my boat, four here at Mukilteo, and two on sheer walls in British Columbia. It had been a fairly long training path of six, cold winter months to attain the trimix certification.

Mukilteo was a boring dive. The giant sea lions liked to zip about a diver in the shallows, and you might spy the occasional fish, skate, or crab, but mostly there was sand, silt, and mud. The advantage to training at Mukilteo was virtually unlimited depth only fifty feet from the parking lot. It was also dark, cold, and the currents in the shallows could be an unpredictable challenge if hanging when a southerly wind started blowing without warning. Although it was definitely not a great place for a sight-seeing dive, the extended underwater time had renewed my respect for the benefits of spending quality time on the bottom and not just bouncing down for a quick peek. The extra fifteen to twenty minutes actually

THE NEW WAY

working in some fashion at depth instilled far greater confidence and proficiency than the two to three minutes on the bottom I had practiced solo.

But what most excited me about the "new" deep diving was the wonder of helium. My sense of orientation and control - despite the vast nothingness of the nearly flat sandy bottom - was different than anything felt before. There was a sticker in Ron's dive shop that read, "Been there, can't remember it...Air. Been there, seen it, remember it...Trimix." It was during my first dive to 240 feet that the significance of the saying finally stuck me like the barb of a spear; the accompanying narcosis was similar to what would be felt at just 115 feet if breathing air. It was incredible, empowering, a feeling that the limits posed in my previous diving life had just been completely blown away.

The simple concept of mixing non-narcotic helium into the breathing gas did, however, bring with it a wide array of technical considerations. We spent many days in the classroom prior to my first training dive with Ron. The one-on-one course was comprehensive: physiology, basic physics, dive planning, equipment configuration, and how to set up the eighth of a ton of gear on one's body for balance, accessibility, streamline, and comfort. These general topic headings were only the beginning of a focused study to ensure a full understanding - at least to the extent that it was understood by medicine and science - of what the pressure and different gas mixtures were doing to one's body. Perhaps most importantly, the true nature of the risk involved in technical diving was directly confronted.

Not long after my certification dive, a trimix instructor named Garrett Weinberg died at Mukilteo following a five-minute sojourn to 300 feet. For an undetermined reason, he surfaced rapidly from sixty feet deep skipping almost all of his decompression. He managed to swim to shore, crawled onto the dock, and began to breathe pure oxygen before becoming incapacitated by the bends. Five hours later - and twenty minutes into his re-pressurization in a hospital recompression chamber - his heart gave out and he died from massive, explosive decompression sickness. The greater depths attainable by utilizing trimix were correspondingly less forgiving of gross error.

SETTING THE HOOK

During my first four training dives we did not even use trimix. These initial dives were a part of the "technical" course, the portion that was focused on how to dive with double, high capacity tanks and the advanced gear that went with them. My past diving experience prepared me well for this topic and the course was abbreviated. I was evidently not the first guy to come back to upgrade my training after extensive deep diving experience, and the instructional agency that awarded Ron his credentials - the International Association of Nitrox and Technical Divers (IANTD) - had a detailed equivalency of experience matrix outlining exactly what steps were required.

I ended up taking two written tests, one for "Advanced Nitrox" and the other for "Deep Air," courses that never existed sixteen years prior. The two tests qualified me to start the advanced technical training, but only after conducting an in depth discussion and a review of my logged dives with Ron. Only then were we able to start on the "Technical," "Normoxic Trimix," and "Trimix" courses. "Normoxic" trimix added helium to the breathing mixture to replace nitrogen, but at a ratio not to exceed 30% helium, and maintained the oxygen content close to that of air. An unqualified "Trimix" certification implied that the amount of oxygen was reduced further to lessen the chance of oxygen toxicity when deeper than 200 feet. The helium content could go as high as to replace all nitrogen (called "heliox"), and the oxygen content could go as low as the equivalent of 16% on the surface after factoring in the effect of partial pressures at depth. I had spent nearly forty hours in the classroom on individual instruction with Ron by my last training dive.

We swam slowly and evenly. It was not the race to the bottom that I was used to, fighting for every second that the old, Navy Dive Tables allowed for bottom time. The Navy's original definition of bottom time started the second a diver's head went underwater, and ended once leaving the bottom at what I came to realize was a frighteningly expeditious ascent rate of sixty feet per minute. The old sixty foot ascent rate had no scientific rationale, but was merely the set variable for the Navy's "Guinea Pig" test divers.

THE NEW WAY

There was now abundant scientific evidence and data to accurately compile reasonably safe dive profiles, and there was new theory supported by the Doppler testing of divers to conclusively determine if bubbles were forming in their blood stream. The data had been around for years, and while there were still unknowns as to the exact hows and whys, the track record of thousands and thousands of deep recreational dives indicated that the new tables were exceedingly safe. There were dozens of computer programs available, some retail, and some - like the one we were using - "freeware."

Why would someone take the time and trouble to develop a decompression program just to give it away? It was difficult to sell such a unique product for a profit in such a small, niche market; there were simply not that many trimix divers in the world. Once the for-profit specter was raised, so too was the probability of liability issues when someone inevitably screwed up. The nitrogen data for decompression was the part that had been tested empirically and was used ubiquitously by all divers breathing air and nitrox. The helium portion, however, was based strictly on a theoretical model extrapolated from the differences in atomic weight between nitrogen and helium, and there existed no defining proof just yet as to its true safety. Still, a sizeable amount of anecdotal data was circulating in the trimix Internet community, and it all seemed to indicate that the helium algorithms were just as safe as the nitrogen model. Individual divers wanted the most user-friendly computer programs possible, and if none existed, they created them. A new breed of techno-adventurers, computer geeks with the stamina to lug around 250 pounds of dive gear, had invaded the sport. Good for them; it was the ultimate and fitting revenge of the nerds.

We could compensate for our slow descent down the sandy slope of Mukilteo by accounting for that factor in our computed dive profile. Ron brought a laptop computer to the dive sites (virtually all serious technical divers did) so that we could analyze the tanks to be used, compare the exact gas mixtures for each diver, take the most restrictive (the combination of the most helium and least oxygen in all buddy team tanks), and "cut" tables real time. The decompression information would then be written down on dive slates to bring with us. It was incredible to realize that gone forever were the days of relying on the set parameters of the old Navy Dive Tables. For each percentage change in the ratio of the "back gas" we would breathe on the bottom, or the decompression gases in each of the two stage tanks hanging under our arms, a new table could be generated in seconds to reflect the slight change to the required decompression profile.

For my final training dive our tables were generated for a gas mixture on the bottom of 15.5% oxygen, 44.6% helium, and the rest nitrogen. During decompression we would first breathe from a tank filled with 34% oxygen and the rest nitrogen, before going to the shallows to breathe from a second tank of pure, 100% oxygen. The exact concentration of each gas, and the appropriate planned depths and times, were calculated to determine the amount of helium and nitrogen that would be saturated in the body at the end of our computed bottom time. All the data regarding how long to spend at a given depth for each of the four tanks and three different gas mixtures was written down on an easy to read curved slate strapped to my left forearm. After finishing the classroom sessions and making a dozen practice dives (some I did on my own), it really was not that complicated or difficult. The evolution of technical diving during the past sixteen years was truly incredible.

We had planned for a sluggish twenty foot per minute descent to compensate for the shallow-angle of the bottom contour. As my depth gauge went past 160 feet deep my bottom timer read eight minutes; twenty feet per minute exactly, right on the mark. The bottom was completely devoid of natural light, and a total blackness instantly filled the beam's abandoned path as I slowly scanned forward. The light-head's oval bracket was designed for a hand to slide through, allowing free use of one's fingers as the beam came over the back of the wrist. The coffee cup sized light-head was connected by a thick electrical wire to a cylindrical battery pack, which in turn was attached to the steel back-plate that served as the frame for double tanks. It was a tremendous help having the use of both sets of fingers while still being able to precisely direct a brighter light than I had ever used before. This was exactly the type of technical innovation that permeated virtually every facet of my new dive equipment, requiring me to start virtually from scratch with almost entirely new gear. Technical diving was not for the financially faint of heart; "buy" the time I was done outfitting, my wallet was over $8,000 lighter.

The steel back-plate, contoured slightly for better fit, was the spine of the technical dive rig. Bolted to the back-plate and weighing in at well over one hundred pounds were twin, steel tanks containing trimix. Attached between the tanks and the back-plate was the rig's "wings" style buoyancy compensatory, with an inflatable bladder that deployed behind my back, allowing for greater freedom of movement. Many wreck divers in the 1970s and 1980s did not use buoyancy compensators at all, finding the jacket style bladder that encircled them too restrictive to movement when inflated. Instead, they relied completely on their dry suits for buoyancy control. In the past, I had put up with the marginal loss of mobility in exchange for the advantage of having the redundancy of two methods of buoyancy control. With both a dry suit and a BC, a safe ascent could still be made if a dry suit sprang a leak (and remember; my old hunk of rubber dry suit used to leak like a sieve). Now, no such compromise need be made. The wings style buoyancy compensator was completely out of the way, yet large enough to provide one hundred pounds of lifting force if necessary. From what I had heard, most of the old dry suit only diehards on the east coast had eventually - maybe even enthusiastically - accepted the innovation.

Balancing the battery canister attached to my back-plate on the right was a fifteen-inch aluminum high-pressure cylinder on the left. This small cylinder carried argon gas for my dry suit. Helium's high rate of heat transfer made it a cold gas undesirable to pump into a dry suit, where argon was denser and felt warmer the second it hit the body. The small argon tank was the final of five that I carried on this typical, technical dive.

I shined my light at the gauge on my left wrist and studied the read out; 240 feet deep. Ron suddenly turned to me and used the pre-briefed signal for me to simulate a complete gas delivery failure. I dropped the regulator out of my mouth in the blackness, swung my light around to my left and behind, and looked for John Campbell to complete the exercise. I had to admit; it was unnerving to have 240 feet of water above my head and to be swimming to a dim light in utter darkness with no regulator in my mouth. That was why it was practiced. With methodical deliberation, I approached John slowly, indicating through body language that I was not panicked, and gave him a slashing sign across my throat with my left hand. He presented his spare regulator, the one attached to a stage tank at his side. I pulled the regulator loose from its position bungeed secure to the tank valve, put it in my mouth, and began to breathe again. The exercise was complete once I resumed breathing off my regulator and re-stowed his spare one.

The simulation was slightly different than most I could expect. John was using a closed circuit re-breather, a complex system that re-circulated his gas mixture, effectively giving him significantly more gas to breathe. There was only one mouthpiece on his re-breather, making it look a bit like a 1950s two hosed regulator, and sharing gas from the re-breather itself was not practical. As an emergency contingency John carried a spare stage tank containing trimix. It was an example of how critical it was to know a buddy's equipment set up in addition to your own. If a gas-sharing emergency was encountered while diving with John, I would need to breathe off the stage bottle on his right side. In a similar situation if I breathed off one of Ron's stage bottles at this depth it would probably kill me. Ron's stage tanks were strictly for decompression at shallower depths, and the higher oxygen contents in them would almost certainly cause convulsions and drowning at 240 feet deep. The Navy flight training mantra "Attention to detail is the key to survival" came to mind.

Similar to my setup, Ron was breathing from a seven foot long regulator hose that was wrapped under his right arm, over his left shoulder, and back around to the right side of his face. His plan in an out of gas situation would be to donate the regulator he was using to me by unwrapping it quickly from behind his head while simultaneously putting his spare regulator in his mouth. Ron's spare regulator hung securely around his neck by a loop of surgical tubing mere inches under his chin. None of the equipment set ups were simple.

THE NEW WAY

Having finished the drill, the three of us turned back to the task of running a second line from the metal spike sunk in the sand at 240 feet to a new one we would place at 260 feet deep. Ron set the line quickly and I occupied the remainder of our ten minutes on the bottom practicing drills. We hung neutrally buoyant four feet off the bottom, careful not to stir up the silt and destroy the artificial visibility created by our light beams, and I removed and replaced each of the stage tanks that were clipped to my harness one to a side. The webbed chest harness kept the stage bottle rigging rings tight and out of view low at my hip and the exercise needed to be accomplished completely by feel. The next simulation was of a primary regulator failure and major gas leak. I began breathing from the spare regulator hanging from my neck and reached behind my head until finding the "faulty" regulator's tank valve. After about a dozen revolutions the valve finally closed, and I hesitated to show Ron that this vital part of the drill was complete before wheeling the plastic knob back open and switching regulators again.

Linking the two tanks together with a common manifold was yet another wreck diving innovation. Unlike my old system with two completely independent tanks and regulators, the common manifold ensured that during normal operation there would be no reason to switch regulators during time on the bottom. The new system's set of three valves on the manifold allowed all the gas in both tanks to be breathed even if one regulator failed. The open water wreck divers had finally opted en masse to use the system cavers had been utilizing for a generation. Each regulator could be isolated by shutting off its respective tank valve. The trick had always been how to reach the valve to shut it down, but with the new back-plate, wings buoyancy compensator, and proper fit of the webbed harness it took only a minor contortion to reach behind one's head to access each tank valve. The valves were downstream of the common connection between the tanks which enabled the entire contents of both tanks to still be accessed even with one valve closed. If a regulator malfunctioned using the old, completely independent system of the 1980s, only half the air would be available. The ability to breathe 100% of a diver's main gas supply in an emergency versus just 50% translated to twice the time to review options and act.

As a final safety feature, a third valve was installed in the middle of the common manifold linking the two tanks. In the event of an uncontrollable leak from a tank valve or a rupture of an overpressure "burst" disc, this isolation valve, albeit with a bit more contortion, could be secured dividing the main breathing supply in half. In this worst case scenario, the diver would be back to being no worse off than we had been in the 1970s and 1980s if a similar situation had been encountered. All of the new gear was great, but it wasn't worth a damn if it was not completely understood and easily reached and utilized, which meant lots of practice until the drills became second nature.

Our timers clicked over in near unison to twenty-three minutes signaling that it was time to start up. I fluttered my feet with a twisting motion to turn toward shore while taking care to not kick up the bottom. Even the common act of re-positioning, of turning around during a dive, had taken practice. In part it was due to the incredible quantity of gear surrounding me, but to be truthful, I had gotten lazy in my diving habits the past several years. Most of my recent dives prior to beginning technical training were spear fishing excursions, and I had picked up the bad habit of becoming negatively buoyant and resting my knees on the bottom while loading the spear gun or waiting for a fish to swim closer. This was not dangerous, merely bad form, at least when doing something as simple as spear fishing. Inside a wreck, or even on the sloping bottom at Mukilteo, touching the bottom meant stirring up a cloud of silt that would destroy visibility in an instant. In addition to negating any reason for being there (why go if you were not able to see a thing?), reduced visibility could be hazardous if trying to exit a ship's hull, return to the anchor line, or in the case of today's dive, to follow a thin nylon line back to the surface. Between the new gear and my erosion of discipline it had taken almost ten dives to regain complete control of my buoyancy and body attitude throughout the changing pressures of varying depths. It had taken longer than expected

THE NEW WAY

for me to re-master buoyancy control and make underwater maneuvering second nature again.

Our ascent rate was a sluggish thirty feet per minute, or a foot every two seconds. Two months ago this seemed aggravatingly slow, but now it felt about right. Helium was a "faster" gas than nitrogen; it dissolved in the body tissues faster, but it also came back out of the blood stream faster as well. Technical divers learned to make "deep stops" to compensate for the faster gas. The trimix community discovered primarily through trial and error that going directly to the computer mandated decompression stops resulted in varying levels of fatigue, which was probably an indication of very slight, sub-clinical cases of decompression sickness. We made our first two minute decompression stop at 160 feet deep, a level far deeper than most recreational divers ever ventured.

We crept up slowly. At 135 feet, my left hand found the tank valve snapped close to my chest and I turned on the stage bottle containing the 34% nitrox mixture. By 125 feet, I had pulled off the surgical tubing wrapped around the regulator's second stage and freed the mouthpiece from its position secured to the tank valve. Reaching 120 feet, the point where a 34% mix of oxygen would equate to a safe partial pressure of 1.6 atmospheres, I dropped my primary regulator from my mouth and began to breathe from the stage bottle. We hovered there for two minutes, and then began our slow progress upward once again. Breathing a mixture higher in oxygen, with no helium, and less nitrogen than existed in air would accelerate the off-gassing of the remaining helium still in the body.

At one hundred feet below the surface, we slowed our ascent further to an excruciating snail's pace of ten feet per minute - that was only a foot every six seconds. By the time we finished two additional minutes at eighty and sixty feet each, we were finally ready for the "required" decompression stops derived by the laptop computer and written on our slates. As my timer cycled past forty minutes, I shivered in the knowledge that our planned profile would have us in the water for an additional hour. We would only be able to surface that quickly due to the 100% oxygen breathed from the second stage bottle hanging under my right arm during the final forty-one minutes hanging at twenty feet.

    I continued to shiver. A new dry suit and "dry gloves" were still on my wish list of gear to buy, but they would push my eight thousand dollar expenditures to over ten, and I was having a tough time adjusting to that financial milestone. I hovered, shivered, and gently swam in circles to keep the blood flowing, wondering how the hell I had gotten here. This was all more difficult than I had expected. This was all more intense than I remembered. I liked it.

CHAPTER EIGHTEEN

A balanced respect for the danger

April 2001, Washington State

The simple challenge of increasing my proficiency - and testing it under the unforgiving conditions of tons of water pressure - became addictive once I began to personalize the new gear, my "rig." There was so much stuff, all still shiny new, yet strangely old and familiar, these futuristic incarnations of my memories of the old way of diving. Tapered lift bags to eliminate the chance of deflation during ascent were now common, and they were smaller, easier to carry, and far more visible on the surface. The local hardware store became a regular stop for surgical tubing and various other eclectic fastening devices to secure my new toys in strategically placed spots around the dive rig. Each piece of accessory gear, like a lift bag, needed to be streamlined, yet easily accessible.

Few pieces of gear stayed permanently where originally placed. Lights, knives, gauges, lift bags, reels; they all migrated through incrementally smaller changes to better locations with each practice dive. I was still making minor changes and adjustments after ten dives using the new equipment, spending many guilty hours in the garage trying different gear arrangements, while fighting the pull of my children's demand for attention in the next room. Enlisting the kid's help in washing dive gear was a pretty weak attempt on my part to stay involved with both.

But the real challenge of trimix diving was perfecting technique. On my first technical training dive, Ron Akeson staked out a triangle of guide lines, each leg sixty feet long, to test my ability and comfort level underwater. We had bypassed the option of making the dive in a heated pool, opting instead for the more realistic environment of the open water, and while the sandy bottom was only twenty feet deep, the water was murky, forty-two degrees, and the current was starting to run.

The first drill was to simply swim the complete course once without interruption to get used to the weight, drag, and placement of the new equipment. A repeat circuit was then completed with mask flooded with seawater, eyes closed, and - just to make it a challenge - swimming the first leg on a single breath of air with regulator trailing behind. Ron's blurry form was waiting for me at the course's first turn point, and I gave him the out of air signal and carefully took hold of the regulator being thrust in my face. It took all my self-control to ignore the icy cold on my face and the urgent pull in my chest, to stay calm and not risk a fumbled grab for the regulator. We then finished the course swimming side-by-side with my mask still flooded and dependent on Ron for both breathing air and sight. The mountain of gear made even underwater swimming seem new and everything seemed just a bit harder, at least initially, than the "way we used to do it."

It seemed that every part of the training was a challenge. On our second set of training dives, we left the dock at the tail end of a rare Whidbey Island blizzard. It took fifteen minutes shoveling snow off the deck before I could even attempt to start the boat, but two dead batteries and a jump-start later we were on our way. The snow competed with the driving wind in a game to see which made choosing a dive spot more difficult, but we finally managed to find a cove relatively protected from the building waves. There was no question in my mind that we press on to complete the day's training; gearing up on a rolling boat deck in adverse conditions was definitely a trick that might come in handy over the *Doria*.

We concentrated practicing most of the same skills over and over, but removing and replacing the stage tanks while swimming at an even pace was the most difficult for me. I rigged one large brass snap to the valve at the top of each stage cylinder with 3/8 inch nylon rope, and placed a second snap on a line running beneath two hose clamps near the base of the tank. It took a lot of initial fumbling before I could unclip and reattach the bulky tanks to the "D" shaped rings sewn into my harness webbing. One D-ring per side was shoulder height at my chest and the other was toward the back of my hips. I must have removed and replaced the stage bottles two dozen times before the awkward contortions while swimming became reasonably second nature. A diver needed to be able to effortlessly reposition or remove stage bottles in the event they became entangled, or in my situation, so that I could tie them off to the wreck before penetrating the *Andrea Doria's* interior. It was pretty clear to me that it would not be a good idea to enter a shipwreck encumbered by the bulky stage cylinders. The stage bottles would not only make it tough to simply swim down a narrow passage, but they could also act as collection points, snagging any line or cable that might be close and making a real dangerous mess out of a penetration dive into the wreck.

I practiced securing and opening the tank valves behind my head to simulate out of gas emergencies, switching regulators with each reposition of the valves until it became second nature. Towards the end of each dive, Ron would signal for me to deploy a lift bag for training. Hovering neutrally buoyant, I would reach back for the lift bag wrapped in surgical tubing at the base of the steel back-plate and unroll it. Holding the lift bag in my left hand, I would blindly use my right hand to find the penetration reel clipped to my crotch strap, also dangling against my butt. It must have looked like I was really trying to pull some wild stunt out of my ass, with each contorted grasp bringing forth a previously concealed contraption.

The next steps in the exercise were to hook the penetration line to the lift bag, release the tension on the reel, and then exhale into the lift bag's narrow opening. The lift bag would then shoot to the surface while I paid out line. The trick was to do it all while maintaining perfectly neutral buoyancy, controlling each inhale and exhale so as not to rise or descend, all to simulate a free hanging decompression away from an anchor line and land. The purpose of the exercise was to provide a free-floating diver both a visual reference point, the reel, during decompression, as well as to provide a surface marker for a tending boat. I worked hard at the drill, fully aware of the seriousness of the situation if blown off the anchor line by the current during decompression. A lift bag on the surface might be the only clue that a diver was quickly drifting away after an *Andrea Doria* dive, and this was a very big deal one hundred nautical miles out to sea in a fog prone area.

The frequent changes to gear configuration did not always make the training dives go easier and there seemed no end in sight to the equipment modifications. For my first dive using trimix, Ron, John, and I crossed the border into British Columbia and drove to Whytecliff Park just north of the city of Vancouver. It was a long walk from the parking lot to the water entry point, but the depth was virtually limitless after a hundred yard swim. We only went to 160 feet (the course of instruction called for increasing the training depths gradually in increments of twenty feet), but a last minute change to the placement of my dry suit-warming argon bottle made the dive far more difficult than it should have been. The small argon tank pressed up against my buoyancy compensator bladder and trapped the gas in one side of the wings. My right side wanted to float; my left wanted to sink. I fought the twisting tendency for well over an hour and my calves cramped up from the constant flutter kick needed to stay upright. To make matters worse, the cotton sweats under my thin, laminated dry suit were not proving adequate for the long decompression and I was freezing. After the dive, simply moving the argon bottle solved the buoyancy imbalance problem, and I ordered a set of synthetic diving undergarments that wicked away sweat and were far warmer. With these problems solved, I concentrated on a bigger issue.

On a decompression dive an immediate swim to the surface is not the best solution to just about any problem. Regardless of how dangerous the situation, shooting up to the sun will almost always make matters worse. Even a simple, relatively protected dive like that at Whytecliff could become a monumental pain in the ass once you were resigned to stay down

A BALANCED RESPECT FOR THE DANGER

for decompression and things went wrong. During the Whytecliff dive, my back was sore from fighting the twisting of the buoyancy compensator, my calves had cramped up, I could barely move my fingers, and my shivering was uncontrollable. But under these conditions simply enduring, "gutting it out," was no big deal. There was sand below me during the long decompression and I could just lie down if need be, there was no current, and the shore was only a hundred yards away. All I really had to do was put up with the discomfort until it was safe to surface.

It was helpful to revisit the concept of gutting it out underwater, of being a prisoner to the ocean until gauges indicated that it was safe to ascend. The ability to rely on reserves to persevere was the ace in the hole for a wreck diver. However, one sure did not want to set out on a dive intending to get through a difficult portion by simply enduring. It paid to leave at least a little stamina in reserve if things really went wrong. Even in the relatively contained environment of Whytecliff Park, if one of my buddies had encountered a significant problem or a strong current had picked up, I would not have been much help while fighting for buoyancy control with fingers numb and shivering cold.

The lesson struck home. It was absolutely necessary to ensure a comfortable gear configuration and to keep warm. Despite the many gear adjustments, I had still been tolerating what appeared to be only minor inconveniences in the arrangement of the new equipment. Whytecliff was a minor epiphany in my understanding of technical diving. I finally honestly understood the fascination, preoccupation, and necessity of perfecting not only technique, but technical gear configuration as well; the equipment was too heavy, ponderous, and complicated to be arranged any way other than perfect.

The road to proficiency was paved with small signs of satisfaction, but by my final trimix training dive at Mukilteo the sense of accomplishment was wearing off. After the twenty-five minute dive to 260 feet, where there was nothing to see but mud, sand, and the occasional stray fish or crab, I would get a shiny new plastic certification card. That was fine, but the reason for all this was to regain that sense of exploration from long ago, before the thrill of the Navy flying and the forced stability of parenthood. It was not anything that could be given to me by someone else. The physical challenge and satisfaction from increasing my proficiency as a diver were not the point. My original goal kept bringing me back around, alternately haunting and fascinating, to dive the *Andrea Doria* again, to recapture the lost feeling of youthful adventure and freedom.

The pieces started coming together once trimix certified, and after months of working on my airline Chief Pilot, I managed to get vacation during the *Wahoo's* first 2001 *Andrea Doria* charter. The boat left port July 2nd, but there was a full week between the *Wahoo's* two planned expeditions for the season. Union seniority issues tied management's hands in many respects when it came to a pilot getting time off. I went through six months of trimix training and $10,000 of equipment purchases not knowing for certain whether I would even be able to get the time off for the dives. It was with a good deal of relief that I learned that the Chief Pilot would rearrange my schedule in order to allow participation in the second expedition as well. Fortunately, he was a diver himself, and must have known that this was not an everyday opportunity.

Steve received my deposits for both trips in March; after all this effort, losing out on a guaranteed spot on the expedition was the last thing to worry about. It had been a busy winter, but the end was in sight and the goal looked achievable. But I hadn't been a great Dad the past few months, and that point was driven home during two brief hospital emergency room visits for separate, and fortunately not ultimately serious, illnesses of my children.

The stress of taking on the project, the dive training, the logistics of putting together the equipment and paying for it, of trying to get the time off, of mentally and physically preparing for the *Doria*, all while flying a full airline schedule were taking a toll. The price was spending less time and focus on my family. But it became gradually easier as each substantial portion of preparation was completed. The training had been mostly grunt work, getting the classroom requirements out of the way, the dives, the equipment configuration. There had been little time to actually enjoy the process, but now the light was at the end of the tunnel, and I was pretty sure it was not the headlamp of an oncoming train.

I kept visualizing the upcoming dives on the *Andrea Doria*, swimming the paths I had finned seventeen years earlier. What was different and what was the same? Captain Janet mentioned in one of her emails that a large portion of the Promenade Deck had fallen off the hull, creating a substantial debris pile at the base of the deck of the sideways sitting *Doria*. The wreck was breaking up with the ravages of time. Would I even recognize the ship?

RICHIE KOHLER NEXT TO THE ANDREA DORIA'S PORT PROPELLER.
PHOTO COURTESY OF BRADLEY SHEARD (1989).

I planned to make three dives per expedition assuming that the weather and currents proved favorable. The excitement of trying to decide where to go on the wreck, what to see, was overwhelming. Eight months prior I would have been satisfied with just touching the *Doria* again, with cruising along her hull and comparing the changes time had made to both us. Now

SETTING THE HOOK

RICHIE KOHLER IS DWARFED BY THE *ANDREA DORIA'S* PORT PROPELLER ON A RARE DAY OF SPECTACULAR VISIBILITY. PHOTO COURTESY OF BRADLEY SHEARD (1989).

A BALANCED RESPECT FOR THE DANGER

that old familiar tug was back, the pull of the artifacts, of going someplace new where few others had ventured. It was difficult reconciling memories of the *Andrea Doria's* lay out with the descriptions of her present condition, but the thrill of exploration was just as remembered; maybe too close to what I remembered.

I needed to be careful. My mind was racing ahead of my experience level. It had been a long time since my dives on the *Andrea Doria* and I needed to fight the urge to make the mistake of the novice, of going a step too far, especially in search of an artifact. I made it a part of my preparation to learn what had been occurring on the *Doria* for the past several years. There were ample web sites to research and plenty of stories to read. More people were dying on the *Andrea Doria* than ever before despite, or perhaps, I thought with a chill, because of the advances in technology. Five divers had perished in the 1998 and 1999 seasons. None of the fatalities had come from the *Wahoo*; in fact, no diver from the *Wahoo* had died on the *Andrea Doria* since John Ormsby in 1985. All five of the recently deceased made their fatal dives from Steve's main competitor, the *"Seeker."* Although maybe no longer an expert in the new style of technical diving, I did know enough not to jump to any conclusions. The sport of technical diving was too new and the cadre of those who attempted to dive the *Andrea Doria* was too small to place cause, much less fault, anywhere without knowing the exact specifics of each instance. But five deaths in two years, regardless of the reason, sure as hell caught my attention. I did not want to become number six.

It was difficult finding a common theme to the accidents. One troubling trait of technical diving encountered during my training and research did, however, provide a possible clue, or at least one contributor to a potentially dangerously unhealthy mindset. It was seductively easy for a diver to become absorbed in the "mission," that overused word thrown around by some technical divers to describe a dive's greater purpose. For the military, a mission is always larger than the individual; there is nearly always a big-picture goal, which often times makes it necessary to run high risks. Exploration for exploration's sake was adventurous and fun, could be educational, and might even on rare occasions have legitimate scientific value, but it seemed that the concept of "mission" had been adopted by the technical diving community in an attempt to create an image of professional stature. Technical diving was still for the most part recreational diving, and to imply that any technical diving "mission" legitimately supported significant risk to achieve some higher calling might have been an attempt to justify that risk to outsiders. I could understand technical diving edging to the precipice of the ultimate sacrifice for the thrill, the challenge, or from a sense of personal accomplishment, but saw no rational reason to actually believe that losing one's life would serve some mission oriented goal. Yet that seemed to be the fundamental implication. Technical diving was undoubtedly dangerous, the appropriate level of acceptable risk was unclear, and the use of a word like "mission" as a pseudo justification did not add value to a widespread understanding of the risk management challenge.

It had always been my experience that successful divers, those who took on reasonable challenges, but avoided death or serious injury, were those with a balanced respect for the dangers of the sport and a clear-headed understanding of their personal motivations. Ultimately, a person went diving and took risks - whether it be underwater or on a mountain - because they wanted to do so as an individual, often in spite of the opinions of others, and not in some convoluted rationalization to legitimize the risk, expense, and effort to outsiders. If a diver actually began to believe that they were engaged in some higher calling worthy, at least in theory, of self sacrifice, then that might potentially dampen common sense survival instincts and make unprepared divers more apt to accept dangerous challenges way out of their league. People often act out of hidden motivations and I saw no value in muddying the psychological waters with lofty terms like "mission."

It seemed appropriate to thoroughly vet such thoughts as my formal training came to an end at Mukilteo during the curriculum's final trimix dive. Ron Akeson hovered to my right twenty feet off the bottom, decompressing on pure oxygen. John Campbell was out of sight to the left, sightseeing in the shallows. My timer clicked over to one hour and forty-two minutes, it was almost time to go up, and I was completely relaxed when shocked out of a near lethargy by a torrent of rushing water. Quickly regaining control, I swam perpendicular to the current, careful not to dip below twenty feet where breathing oxygen could kill, deflated my buoyancy compensator, and dragged my fingers through the sandy bottom until I found an embedded rock and held on. I had been warned of this happening in the shallows of Mukilteo. The waters stretched unobstructed past Seattle to the south, and when the occasionally strong southerly wind picked up it funneled along the coast, causing the current to go from nil to extreme in an instant.

We clung to the rocks for another five minutes before inching up the boat ramp to where we could stand. With the decompression over, the only remaining challenge was getting out of the water. Ordinarily, we would unclip the stage tanks and leave them in the shallows while we clambered under the load of the doubles and weight belts to shore. Only after being relieved of the 150 pounds strapped to our backs would we wade back in to retrieve the additional hundred pounds of stage bottles. The normal plan was clearly not going to work this time; the building waves would drag the stage bottles off in an instant.

I kept my regulator in my mouth, removed my fins, stood up, and started walking toward the beach in four feet of water. Any shallower and I could not have straightened my legs under the weight. With both stage bottles hanging from my harness, I walked the fifty feet to shore under the load of 250 pounds of dive gear. I reached the pier, now completely exhausted, lowered the tanks to the cement, caught my breath and slowly unstrapped and unsnapped the mountain of equipment surrounding me. With dual physical and mental relief, I realized that getting out of the gear was the final tangible, training obstacles between the *Andrea Doria* and me.

CHAPTER NINETEEN

Bragging rights and the twin hazards

Why do adventurers climb great mountains or dive to dangerous depths? The various answers seem to inevitably evolve into platitudes, even the most famous and thought provoking response of "Because it's there." Everest climber George Mallory's elegantly simple acceptance of the inexplicable nature of the answer is not satisfying, nor do I expect that it was meant to be. Why dive the *Andrea Doria*; why bother? Years back, a simple answer for me would have been "To get china," as if the muddy dishes possessed some sort of intrinsic spiritual value. All those basement artifacts were interesting trophies to be certain, and a wreck's history was written in these brass, glass, and ceramic conversation starters. *Doria* china was also tangible proof that at a specific point in a person's life they were skilled divers with either sufficient nerve or stupidity to go into the Dish Hole and come back alive, but being able to identify which of these two choices was the correct characterization was not particularly clear cut either. Then there is the "purpose" explanation, the proposal that in modern times it is less of a challenge to simply survive in western society, and new challenges are needed to lend a person a sense of unique identity, of being special and belonging to an exclusive club. So was that really it, did the motivations for extreme sports simply come down to bragging rights and rebellion against an indifferent society's glacially certain movement toward universal anonymity? Uncomfortably, I would have to answer

"Yes." But nothing is that simple, and there is still a missing element to this passion that defies definition, maybe because it is such an integral part of one's being that it can't be pinpointed. Oh, the hell with it: "Because it's there."

Being quite certain that I had neither the nerve nor the ability to penetrate deep into the wreck any longer, was my plan to scrounge for china in the safer waters outside the wreck? If so, what kind of trophy artifacts were those to be considered? In the distorted reality that comes from familiarity and obsession, it seemed that diving the *Doria* was now common place, even pedestrian, at least when compared to my memories. So what was the point in diving the wreck again? Was it reasonable to expect that I could simply pick up where I had left off with life as a college kid? Of course not, but that didn't mean I shouldn't try. I was looking for a big answer, and all this dancing around the edges was not getting me any closer, in fact, my growing hesitancy was evidence that there indeed was something to be learned by returning to *Andrea Doria*; I was just increasingly unsure of what would be discovered.

Sport diving had taken a giant leap forward in gear, technique, and skills. It was easy to sense the accompanying standards of risk-taking changing as well, a spirited cadence in tandem with the acceleration of technology in the sport. Two hundred feet used to be deep; it was certainly far deeper than any instructor brought a student in the early 1980s. I had just finished my final training dive to 260 feet, a depth great enough to become "certified," but really only a license to learn, to practice and hone skills that could never truly be perfected. What was deep now? What were the new challenges that had supplanted the old?

The limits of old had been complex and vague, but they were limits set primarily by the twin hazards of nitrogen narcosis and decompression. You could only stay underwater for so long - the eighty cubic foot supply of air in each of the twin tanks severely limited the time available on a wreck. A total of 160 cubic feet of "back-air" was small by today's standards, even before factoring in that a significant portion of the air had to be saved for decompression. These depth limits were pushed well out by 120 cubic

BRAGGING RIGHTS AND THE TWIN HAZARDS

foot tanks used exclusively for "back-gas" during the deep portions of a dive, which still allowed for a sizable reserve supply, and plenty of gas in stage bottles to use solely for accelerated decompression. There was now a tremendous safety margin in all facets of decompression diving assuming that everything functioned as advertised, including a diver's judgment. The end result was that the new reality was not that different from the challenge of old; the fundamental risks were the same, the goal posts had just been moved further down the field. In one respect the new risk was even greater because of a diver's dependence on their stage bottles or their buddy for survival. There were a lot more moving parts that all had to integrate and work together in order to be successful.

During my last dives to the *Andrea Doria*, I routinely surfaced with barely any air left in my tanks, virtually no fall backs but quick thinking and my buddy (presuming he or she were around) if something went horribly wrong. With the newer high capacity tanks, we could make longer dives to deeper depths and not infringe on the "rule of thirds." Cave divers used this rule religiously in the 1970s and 1980s. I was aware of the concept, but routinely ignored it twenty years ago in my brash confidence. Not anymore. I now planned on using no more than a third of my gas supply exploring out into a wreck. A third of the supply would be saved for the return to the anchor line and my ascent until I reached a safe depth to breathe from one of the nitrox-filled stage bottles. The final third was kept in reserve for contingencies. By utilizing eighty cubic foot tanks for my decompression mixtures of nitrox and oxygen, I had ample reserve gas for most extended, unforeseen decompression requirements.

The longer time spent at depth brought with it an increasingly catastrophic outcome if one was forced to surface for any reason without decompressing. On the other hand, dead was dead - surfacing after even fifteen minutes at 200 feet without decompressing would probably be fatal without an expeditious re-pressurization, be it in a chamber or back underwater. Nitrogen narcosis could be eliminated for all practical purposes by substituting a sufficient supply of helium in lieu of nitrogen: the deeper the dive, the more helium. The two factors that had previously most

SETTING THE HOOK

egregiously limited our plans on the *Andrea Doria* had been either eradicated or generously relaxed.

But there was no free lunch. A diver's ability to move within the restrictions of the mountains of gear made a penetration of a wreck with stage bottles very dangerous. To penetrate a wreck meant leaving the stage bottles somewhere outside, but where? Through email exchanges with Captain Janet I learned that the preferred method was to unclip the two stage tanks, tie them to a part of the wreck, and retrieve them on the way back to the anchor line. Without those stage bottles, decompression after a trimix dive was next to impossible: the bends were certain, death was likely, and a trimix diver simply had to make it back to his or her stage cylinders or there would be a big problem. I shuddered to think of all the ways that the tanks could become inaccessible. What if they could not be found in the murky outline of the shipwreck? What if they collapsed in a deterioration of the hull? What if they just weren't there, if they had worked themselves loose in the current and surge? Finding the anchor line was important - otherwise a free-hanging diver might surface after a lengthy decompression out of sight of the *Wahoo* and a hundred miles from the Atlantic coastline. But if this happened, at least the diver would be alive when he or she surfaced. Not finding one's stage bottles, especially if diving solo, probably meant death.

Given this caveat, the past limitations to the exploration of the *Andrea Doria* had been largely lifted. I spent months visualizing the wreck, trying to decide what to explore, what to accomplish. There was no way of knowing for certain where on the nearly 700 foot long *Andrea Doria* the *Wahoo* crew would set the hook. Much of what I remembered physically about the wreck might not even exist. But that was what it was all about, wasn't it? That was the point, to go back and see for myself, and to actually see myself, and learn how time had changed us both.

I was definitely more cautious these days than in my early twenties. Perhaps it was my Navy flying and combat experience, or maybe it was my kids, or getting older, but I gave a hell of a lot more thought to safety than when twenty-one. I had ignored the pull of the *Andrea Doria* for seventeen

BRAGGING RIGHTS AND THE TWIN HAZARDS

Harold Moyers swimming with stage bottles illuminates lost fishing nets as he explores the *Andrea Doria*'s stern in 2006.
Photo courtesy of Bradley Sheard.

years, but other divers most definitely had not. Ten divers had died on the wreck since 1985, bringing the total fatalities to twelve since divers began regularly scheduled charters in 1981.

The number of expeditions to the liner had grown in a predictable pattern since Peter Gimbel showed off that first piece of china to the cameras

SETTING THE HOOK

in 1981. Charter boats came and went, some changed owners, and some decided the *Doria* was simply not worth the headache. It was not easy to screen the trove of artifact hunters for what had come to be termed, "china fever," from the ones that were truly qualified. Even if this could be successfully accomplished, there was no guarantee that a smooth talking diver would not do something completely unexpected once in the water and screw it up anyway.

Steve kept the *Wahoo* to its schedule of two, occasionally three, expeditions a year to the *Andrea Doria*. He felt that this was the maximum number of charters that could fill the *Wahoo* with qualified divers; there were simply not that many people around who were willing and able to dive the *Doria* with any degree of safety. Steve's old nemesis Sal Arena, the captain of the *Sea Hunter*, continued to take regular charters to the *Doria* throughout the early 1990s. The first boat named the "*Seeker*" entered the scene in the mid-1980s, bringing to three the number of regularly scheduled charters that ran to the shipwreck.

Gary Gentile explored the *Doria* with regularity over the years, crewing on the *Wahoo* until he had a falling out with Captain Janet and then Steve, after which he went on to crew with the *Seeker*. Gary's friend, Bill Nagle - who I knew in passing from several of the *Wahoo's Andrea Doria* trips - had bought the *Seeker* to run wreck charters. This was the original "*Seeker*," the first of two wreck diving boats out of New Jersey with the same name. Nagle's initial objective with the *Seeker* was not motivated by profit, but from a driving desire to retrieve one of the *Andrea Doria's* singularly unique prizes, the ship's bell.

After a fruitless search for the *Andrea Doria's* bow bell, including a sojourn along the sand at 240 feet breathing air, the Seeker's expedition transitioned to a search of the ship's stern for the auxiliary bell. The group was successful in 1985, and spring-boarding from this early victory Bill Nagle purchased a boat of sufficient size to rival the *Wahoo* in the off shore dive charter business. The new sixty-five foot *Seeker* started operating in 1987. Unfortunately, Bill Nagle was experiencing much deeper problems than many casual acquaintances like me would have guessed. In 1993 he died from alcoholism. After a short interlude, Dan Crowell purchased the *Seeker* in 1995, and the *Andrea Doria* charters resumed in earnest. For fifteen years, rarely in friendly circumstances, more often with casual disregard, and occasionally with visceral antagonism, the *Wahoo*, the *Sea Hunter*, and the *Seeker* vied for position over and inside the wreck of the *Andrea Doria*.

In 1988, the *Wahoo* experienced firsthand one of the potential disasters awaiting divers; a lost diver on the surface. Captain Janet had graduated from the minor leagues of the charter boat world and was conducting the July charter to the *Doria* solo while Steve stayed home. The responsibility put upon her was not misplaced.

Billy Campbell, the same gent who placed the 25th anniversary plaque on the *Andrea Doria* with Steve and Hank Keatts in 1981, had just finished his required decompression hanging on the *Wahoo's* anchor line in a strong current. Billy let go of the line ahead of his buddy, Dave Zubic, and allowed the current to push him back toward the *Wahoo's* stern. Billy simultaneously unclipped his camera and strobe light while allowing the current to glide him effortlessly aft until he could grab the ladder rungs with his camera gear ready to hand up to the crew. Underestimating the current, when Billy next looked up he realized that the swiftly moving waters had pushed him past the ladder and behind the *Wahoo*. He surfaced to get oriented, kicked into the current, and tried to wave to Hook who was on the aft deck of the *Wahoo* briefly looking down while he filled a dive tank from the nearby air compressor. The *Wahoo* was only about thirty feet away, but she was surrounded by a thick fog bank obscuring visibility to virtually nil.

Billy yelled to Hook, but his voice was lost in the loud chatter of the air compressor. He took a compass sighting toward the dive platform and descended to ten feet underwater where he could better fight the drag of the current against his dive gear. Still believing that he could make headway against the swift waters, he opted for the direct course to the boat instead of angling his swim toward the buoyed safety line strung out behind the *Wahoo*. After a minute of fighting the current, Billy Campbell surfaced again to check his progress and see how much distance remained until reaching the *Wahoo's* dive platform. All he could see was white, misty infinity.

SETTING THE HOOK

Billy Campbell was alone, one hundred miles from shore in the Atlantic Ocean, in a current, in the fog. The utter isolation of mind and body that had touched so many connected with the *Andrea Doria* took hold of Billy Campbell. It was a shared experience, a trip to the *Andrea Doria* threatened by disaster, followed by a profound disconnect from the rest of the world.

Captain Piero Calamai had shared this experience. Unlike his ship's namesake, Admiral Andrea Doria, who earned fame through victory over the French and the Barbary pirates, Calamai's fame was of the ignominious and lonely sort. He never recovered from the *Andrea Doria's* sinking, and with reputation tarnished forever, he would never truly live in his old world again. His was a lonely position to start; the collision sealed his fate as a recluse to tragedy for the remainder of his days.

Martha Peterson, trapped beneath the rubble in her cabin, must have felt the isolation until her dying moment. The confusion of being thrust into a situation she could neither control nor fully comprehend was only dampened by the tireless efforts of her husband's rescue attempts. Although her husband was at her side when she drew her last breathe, he was free to escape, while she became a guest of the *Andrea Doria* forever.

Flung onto the bow of the *Stockholm*, Linda Morgan was totally alone as the only *Andrea Doria* passenger on board the *Stockholm*, immediately removed both physically and mentally from her accepted reality.

Robert Hudson, awakening in the darkened, tilted world of the *Andrea Doria* during her final death heel and alienated from the rest of the world, struggling like so many divers would in the years to come to escape to open air and survival.

And young Norma di Sandro must have sensed the aloneness in some primal way, parted from her parents by her father's good intentions in trying to save her, by dropping the child over the *Andrea Doria's* Promenade Deck rail. She never left a coma during her brief reunion in this world with her mother and father.

Frank Kennedy knew the isolation, both from within the depths of the wreck and on his lonely claw to the surface fighting with futility for survival.

BRAGGING RIGHTS AND THE TWIN HAZARDS

Sally Wahrmann, deep in the pitch-blackness of the wreck with 252 feet of water and steel above her, mask flooded and regulator floating behind her, was well aware of the experience too.

Most divers on the wreck had felt the separation, be it only for a few minutes in the silty mess of the Dish Hole, or for eternity as a captive to her cables or passages. All who explored the *Andrea Doria* shared a touch of common experience with her passengers, the excitement of survival, of not knowing what fate would reveal while the liner's guest.

I had felt it intensely those terrifying few minutes in 1983 when silted out blind and lost, free-floating 210 feet deep in the First Class Dining Room.

And now, just when he must have felt the seductive urge to relax in the illusion that he had cheated the *Doria* out of a shared company for all time, Billy Campbell was completely alone in the white mist of collision between the Labrador Current and the Gulf Stream. At times it seemed that fate was using the *Andrea Doria* as a vehicle to divide and conquer, to isolate an individual until they profoundly lost their footing, to see if they could find their mettle in a one on one mind game of will and physical perseverance.

Billy's dive buddy, Dave Zubic, surfaced and spread the alarm minutes after Billy disappeared into the fog. The question sat hanging for several minutes, what course of action should be taken, how could the *Wahoo* search for Billy and still recover the divers underwater on the wreck? Captain Janet knew Billy was almost certainly on the surface by this point, moving away from the boat dangerously quickly and unable to signal his whereabouts. Janet's first action was to call the Coast Guard. It takes time to get a helicopter prepped and into the air, and much longer for a cutter to travel the gap to the *Doria* from New York, Rhode Island or Massachusetts. The wheels needed to start turning on these time consuming evolutions. It became clear that the *Wahoo* could not pull anchor and search the fog while divers remained on the decompression line. If the anchor line was buoyed and released, the decompressing divers would be in almost as precarious a position as Billy when they surfaced to find no boat and were forced to continue clinging to the line. The chase boat was the obvious answer. But the *Wahoo* had radar; the inflatable chase boat did not. How were the two to rejoin if the chase boat ventured out into the mist in a search pattern?

Captain Janet and Gary Gilligan did some quick thinking. First they raided Sally Wahrmann's galley, where she heroically donated a roll of tin foil and a package of Styrofoam cups (Sally was rather protective of her kitchen). They then rigged the end of a broomstick with tin foil and made an ad hoc radar reflector. Armed with this locator for the *Wahoo's* radar and a handheld two-way VHF radio, Gary and Dave Zubic began to slowly motor the inflatable deep into the fog behind the *Wahoo*. Gary Gilligan did one other very clever thing; taking a stack of Styrofoam cups, he began releasing them into the current in steady intervals every few minutes. The pair of rescuers utilized the floating cups in the limited visibility as a reference to the current's direction in the hope that they would point to Billy.

By listening to the momentary blasts of the *Wahoo's* foghorn and following the compass headings transmitted over the radio by Captain Janet, Gary was able to precariously maintain contact with the *Wahoo*. Gary motored slowly, following the current and the Styrofoam cups as best he could, aware that Billy was floating further away by the second. The chase boat was on a tenuous tether, and once the jerry-rigged radar reflector grew too faint on the *Wahoo's* radar, Janet would call on the radio and instruct them to move closer. Two additional people lost in the fog would have complicated the search problem significantly.

After more than an hour the chase boat returned with its frustrated pair of searchers. Meanwhile, floating far behind the *Wahoo*, Billy took stock of his position. In any survival situation the difference between life and death is ultimately dependent upon the quality of the initial decisions. A tired body leads to a tired mind, and a tired mind leads to poor decisions. With no idea how long he would be floating in fog bound oblivion, or if the fog would even lift in time to rescue him, Billy realized that he needed to conserve his strength. He was wearing a diving dry suit, which played greatly in his favor. But the buoyancy would only help as long as he could hold his head up out of the water, and sleeping could prove fatal. But the dry suit did mean that at least he could stay warm enough to survive almost indefinitely. He must have wondered how the sharks would react to the bobbing flotsam of his body. With admirable discipline he decided to jettison all the dive gear that was of no direct use to him. It would cost him thousands of dollars, but by ditching it he could conserve vital energy. Billy did keep one piece of his equipment, not for sentimental reasons, but strictly for survival. He held on to the powerful underwater strobe and his video camera. He needed something to use as a signal if he was still adrift when darkness fell. Billy also wanted a way to save his final thoughts for his family if not found alive.

It was abundantly clear on the *Wahoo* that a narrow window of opportunity to rescue Billy was closing, and if Billy were to survive the search would need to resume immediately. Gary Gilligan set out again in the inflatable and followed the bobbing Styrofoam cups in the current. Two and one half-hours after Janet's initial radio call, the sound of rotors beating the air into submission could be heard overhead as the Coast Guard helicopter went into a hover, but the helo could offer no help until there was a break in the fog to allow an airborne search. If Billy Campbell was still within the radar range tether of the aluminum foil wrapped broomstick on the chase boat, he would not be for very much longer.

Three hours into the ordeal, at the outer range of Janet's ability to identify the broomstick's radar return, the chase boat found Billy Campbell. He was three and a half miles behind the *Wahoo* and drifting fast. Radar had saved the day over the same spot where it had proved so ineffective in preventing catastrophe in 1956.

Diving the *Andrea Doria* was a lonely experience. Even with the best dive buddy in the world, one was still ultimately operating solo. As the diving manual from my trimix class had so correctly pointed out: "Nobody can breathe for you, swim for you, or think for you." There would be no verbal communication, no talking while on the wreck to reassure and scare the ghosts from a diver's mind. The popping of a diver's expanding bubbles in a race to the surface was the only sure sound. And regardless of how well a dive was planned and briefed, there was always a very good chance that the *Doria* would capriciously decide to isolate and test a diver anyway. I could not help but wonder if fate's work was made easier as I completed final preparations for a return to the Andrea Doria - alone and without a buddy.

CHAPTER TWENTY

"Simple answers ain't always so simple..."

Over the course of the twenty years since Peter Gimbel's 1981 expedition, the allure of *Doria* china did not lose a sparkle of luster to the divers who got a peek at the handcrafted dinnerware. The number of explorers willing to risk it all for a piece of the ornate ceramic grew in unfortunate proportion to the disappearance of the more accessible dishes. Most first timers to the wreck were content to touch her, to be able to say that they "Dove the *Doria*." On my first successful *Andrea Doria* expedition in 1983 several divers new to the boat had sported T-shirts emblazoned with the words "Happy are those who dream dreams, and dare to make them come true...*Andrea Doria* dive 1983."

Craig, Gary, Don, and I liked the catchy saying, but thought the melodramatic tone a bit over the top. We had one of our own "*Wahoo*" T-shirts modified slightly: "Happy are those who dream dreams, and dare to make them come true...Wally's Pub." Wally's was a favorite local watering hole, and there was little danger of anyone thinking of him or his establishment in solemn terms. Oddly enough, come 2001 this was the only *Andrea Doria* T-shirt (out of an initial stock of a dozen) still in my possession, and sadly, Wally's was out of business. If a diver did not take themselves too seriously, perhaps it would dampen their appraisal of personal ability and help them avoid overextending their limits. Confidence was important on a dive, but a bit of humility and a sense of humor could be healthy too.

SETTING THE HOOK

Still, for most it did not take long to succumb to "china fever." Maybe it was simply the spirit of exploration, of overcoming a challenge; in many divers there was little doubt that was the answer. In others it was something else. Call it greed, pride, competition, or bragging rights, whatever it was it proved to be deadly. One of the hardest things in the world is to go to phenomenal lengths to reach a goal, and then to have the maturity to accept retreat once a pre-determined limit has been reached, especially when there are no other voices to rein in a diver. "Plan your dive, dive your plan" has always been the clichéd mantra for a safe underwater adventure. It should be easy to turn around when the planned limit has been reached: the minimum gas pressure, or bottom time, or maximum distance of penetration into a wreck, right? Bullshit. When a diver knows it's their last chance (maybe only for the season, maybe for real), when the fear of being aced out of a prize darkens the subconscious, the urge to cheat on the plan just a bit, to penetrate the wreck just a bit further, to spend just a few extra minutes, breathe down the tanks a touch more becomes nearly impossible to ignore. Only one person can make the decision, and it must be made in complete solitude under cover of water, steel, and darkness.

"Simple answers ain't always so simple, even a dumb Polack like me can see that," Steve Bielenda liked to say.

Most of the fatalities on the *Doria* came from those relatively new to her. Maybe the "old timers" had figured out this basic rule early on, or they had simply accumulated sufficient experience to realize intuitively that one first needed to be alive tomorrow for another opportunity to present itself. True *Doria* aficionados like Gary Gentile, John Moyer, and Gary Gilligan would never be satisfied with "just" china; they were clearly bonded to the Italia liner for the long haul.

I had seen both ends of the spectrum. My unplanned stray into the First Class Dining Room had almost cost me everything. But I had other opportunities where circumstances worked out differently, yet were still monumental tests at the time.

"SIMPLE ANSWERS AIN'T ALWAYS SO SIMPLE..."

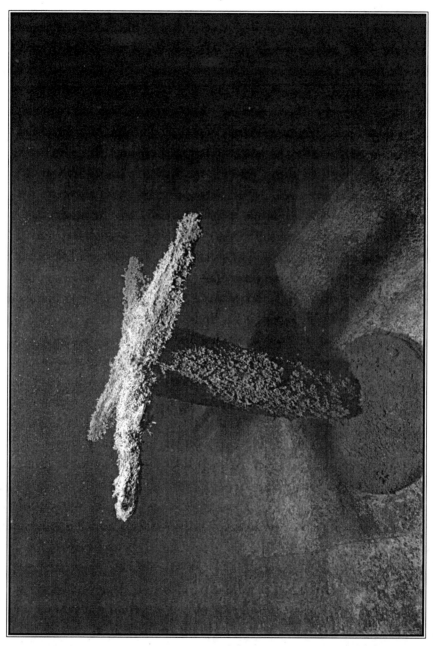

Table minus top in the First Class Dining Room where the author became lost in the silt in 1983 (photo from 1988). Photo courtesy of Bradley Sheard.

SETTING THE HOOK

The year after desperately seeking an escape route from the *Andrea Doria's* First Class Dining Room, I made a dive to the *U-853* with my friend "Matt the Cat." Matty would very seriously explain to people that he was from the Bronx, a pat and comprehensive answer to all his eccentricities and off the wall antics. He worked for the New York Transit Authority, where he proudly "kept the wheels turning," single handedly in fact, just ask him. His straight-faced demeanor barely revealed the hint of a cockeyed grin below his mustache when he was cooking up a scheme. But whether he was boiling lobster with a welding torch in the back of a transit bus up on jacks, or writing checks for wreck artifacts from third parties to keep up with the rest of us (Matty had no desire to attempt the *Doria*, he might fool those around him, but never himself), Matt the Cat was lively, entertaining, and fun. He was the consummate bullshitter, our "Checkbook Diver."

That was why it was so incredibly difficult for us to know when to leave the hull of the *U-853*. Matt was a good diver, not as experienced as the rest of us, but well aware of his limits. While finning over the top of the U-boat he came across a cargo-loading hatch, and intrigued with the prospect of actually recovering an artifact without having to break out his checkbook, he enthusiastically got to work. I tried to help, but this was Matty's show. After fifteen minutes, using vise grips, a screwdriver, a drift pin, and a five-pound hammer (Matty always carried a lot of tools, he was a mechanic after all), the hatch was free. We had already rigged a lift bag ready to be inflated, and it would only take one to two more minutes of time and air to get it to the surface.

Unfortunately, we found the hatch late in our dive. We were at only 100 feet deep, but our tanks were almost empty and we required a minimum of eighteen minutes of decompression according to the old Navy Tables. We definitely did not have enough air to fill that lift bag, probably not even enough time to fill the bag if we did have the air. I tapped Matty on the shoulder and gave him the thumbs-up sign; it was time, way past time really, to start our ascent. Matt the Cat looked at me in disbelief, shook his head, and turned back to work. Forcefully I pulled his elbow, turned him to me, looked in his eyes and gave him the ascent signal again. His body seemed

"SIMPLE ANSWERS AIN'T ALWAYS SO SIMPLE..."

to visibly deflate slightly as he shook his head, took a last look at the prize, and turned to follow me. We swam to the anchor line and started up.

With only one minute remaining in our ten foot hang and with virtually empty tanks, another pair of divers came up the anchor line past us with their own lift bag slowly raising the hatch cover - our hatch cover.

Matty was furious. The divers who surfaced with the hatch did not carry any of the tools required to remove the corroded pins that held it secure to the U-boat. It would have been impossible for them to free the hatch had it not been for our efforts of minutes earlier. Suffice it to say that the argument on board the *Wahoo* was heated, almost to the point of blows. It became so heated that Steve eventually threw the round hunk of steel over the side while the *Wahoo* was underway - no one would keep the hatch. It was a pretty shitty day over all, but we were safely on board the *Wahoo* after the experience and not in a Coast Guard helo on the way to a recompression chamber or worse. We had made the right decision in leaving the hatch. There was one saving grace, however; every beer shared with Matty from that day forth would be prefaced with a loud toast of "Down the hatch." The pained look on the Cat's face was worth the entire ordeal.

The allure of greater prizes was nothing to take lightly, and this dangerously misplaced quest for some kind of glory, a fleeting fame which at best would probably only be truly appreciated by the relatively small circle of *Andrea Doria* divers, stood poised ready to kill. Nature took its toll on the *Doria's* massive hull with the passage of years. Where in 1981 only welding torches could breach the doors to her interior, by the early 1990s corrosion and winter storms had provided access to other areas. But the prime artifact of choice remained the same: china.

The ornate Asian figures of Richard Ginore's artwork made the First Class china the most desirable and these pieces quickly disappeared. But there were two other major dining rooms on the *Andrea Doria* for her passengers: one for Cabin (Second) Class, and one for Third Class passengers. In fact, the same reddish and gold inlaid china that graced the First Class Dining Room along with Richard Ginore's hand painted pieces could also be found in the Cabin Class section.

Standing: Craig, the author holding U-boat mid-ship torpedo loading hatch, Matt "The Cat" Galcik. Kneeling: Don "Hook" Schnell and Tim Nargi. From the author's collection, photographer unknown (1983).

The depths were comparable (210-220 feet), the hole to gain access narrower, and the competition as vicious as ever. Soon after the discovery of the new treasure trove a group of divers from Bill Nagle's *Seeker* decided to gain an edge on the competition. They used torches to cut a wider hole for access to the spot, but did not leave it unattended for other divers to use. When the *Seeker* steamed from the site they left behind a steel grate, padlocked for good measure to keep out the undesirables (i.e. the *Wahoo* divers). Attached to the grate was a hand written sign:

"SIMPLE ANSWERS AIN'T ALWAYS SO SIMPLE..."

CLOSED FOR INVENTORY
PLEASE USE ALTERNATE
ENTRANCE
THANK YOU
CREW AND PATRONS OF
SEEKER

A pretty good one overall - the grate didn't last long.

The Third Class china was rimmed in blue, and was just as beautiful as the other pieces in its own way. The fundamental fact was that anyone, on any dive to the *Andrea Doria* was pretty damned excited to recover a single piece of china. But there were those who would not be satisfied - why should they be when the nearly 700 foot *Andrea Doria* had so much more to offer - and continued with far grander projects plans.

Once the *Andrea Doria's* stern bell was recovered in 1985, the question lingered, what happened to the bell on the bow? A bell had been found years earlier in New England with *"Andrea Doria"* engraved on it, and it was doubtful that it was a fake. But it had a large crack running down its side, and it would have been a strange item to remove from a sinking ship. It could have been removed earlier for repairs, but the *Doria* was en route to New York, how would it have gotten there prior to her arrival? Had the *Andrea Doria* made her last passage to Italy and now back again to New York without a bow bell? John Moyer did some research and came to several conclusions. First, the *Andrea Doria's* primary bell was only presented topside on the bow when the liner was in port. When the ship put to sea the bell would be removed and stored in the bow storage hold, the "paint locker." This would explain why it could not be found during Bill Nagle's 1985 expedition: it was not there. No one had looked in the paint locker though.

John Moyer chartered the *Wahoo* several times in the early 1990s to search for the missing bell. Digging through the chemical tainted silt amidst the corroded paint cans was certainly dirty and dangerous work. His group utilized an airlift, a surface powered, elongated tube that sucked the muck out in a vacuum and discharged it outside the compartment.

Relatively quickly, however, several divers reported becoming sick as a result of artifact hunting in the new spot. And to top it off, the bell was not to be found. John Moyer realized that his search was becoming futile and perhaps unhealthy in such close proximity to the old paints and chemicals, and he transitioned to a more realistic goal. The mystery of the bow bell would yield no new evidence for the next seventeen years until 2010, when what was probably the Crow's Nest bell, but could possibly have been a temporary replacement for the bow bell, was found by a pair of divers in the sand.

The *Andrea Doria's* Promenade Deck curved around the topside of her bow in a magnificently enclosed Winter Garden. Records were clear that certain pieces of artwork had been displayed in this spot, a location that was not particularly deep (as far as the *Doria* was concerned, at about 190 feet), and not a terribly far penetration. Both these factors were critical if the three segments of the 700 pound Gambone mosaic panels were to be removed. John Moyer's research and previous exploration of the wreck made this an attainable feat, and for all intents and purposes he was the first to recover the panels in their entirety. But John Moyer was not quite the first to bring a piece of the Gambone treasure to the surface, as he was to find out later much to his surprise. The three pieces of the mosaic were eventually recovered, but only after numerous backbreaking dives. First, the concrete-backed artwork had to be knocked free from the bulkhead in the Winter Garden, and then carefully rigged lift bags were used to avoid jamming into the ceiling. A method still had to be devised to lift the 700 pounds of stone out of the water while a hundred miles out in the Atlantic. The *Wahoo* had removed its lifting davit to make greater space for divers entering the water several years earlier.

Cleverly utilizing a series of levers, Steve and the group managed to hoist each of the fresco's pieces onto the *Wahoo's* dive platform where they could be secured for the long passage home. The Gambone panels were far and away the biggest prizes to come from inside the *Doria* since Peter Gimbel retrieved the Bank safe in 1981. The only larger artifact recovered was the giant Promenade Deck window frame Steve, Hook, and Gary

"SIMPLE ANSWERS AIN'T ALWAYS SO SIMPLE..."

Gilligan had brought up in 1984, but that had been sitting on the outside of the wreck and was considerably easier to bring to the surface.

When John Moyer set out to recover the artwork he did not want other potential salvors interfering with his efforts, either underwater or through legal means after the recovery. Technically, the *Andrea Doria's* insurance company still owned the salvage rights to her hull. Under the "Law of Finds," repeated and well-publicized salvage attempts over the years - which raised no objections from the insurance company - made a compelling case for an abandoned wreck. But that begged the question: who was in de-facto possession of the salvage rights to the *Andrea Doria* and could legally lay claim to her prizes?

In 1993 John Moyer went to court and was granted an injunction to salvage the *Andrea Doria* and legally keep whatever artifacts he recovered. The fact that Moyer had been working the wreck actively over the course of ten years, and no others had come to court to receive official sanction first, were sufficient arguments to make him the legal owner. Fortunately for the other *Doria* divers, Moyer did not make an issue of his ownership of the salvage rights, and he made no attempt to stop others from retrieving the more mundane (and small) artifacts. China hunters were certainly in the clear.

It was an interesting twist to the artifact controversy. Did possessing the salvage rights to a vessel make any moral difference as to what could be ethically taken? I had never considered this nuance, primarily because of my pronounced leaning toward bringing to the surface pretty much anything shy of human remains. The massive exposure, and some would say exploitation, of the *Titanic* altered the argument from an eclectic diving discussion with an occasional foray into the courtroom to an issue of some international importance.

The Federal Government, on the other hand, had been paying attention, and in 1988 the "Abandoned Shipwreck Act" was enacted. From that time forward individual states held exclusive rights to all abandoned wrecks within their borders (i.e. inland waters) and up to three miles out to sea (nine miles for Florida and Texas). State law would apply in every instance, and you could rest assured they would heavily restrict or tax anything recovered.

For fresh water wrecks and those hulks close to shore, the taking of artifacts were often severely restricted after 1988. But the New York wrecks most frequented by divers were all further than three miles out to sea and were unaffected by the law. The only other common statute applied only to U.S. Naval vessels, which never relinquished the rights to sovereign ships and their content. The Navy enforced their rights with discrimination, however, as witnessed by the recovery of artifacts from the U.S.S. *San Diego* over the course of more than eighty years. In one notable case, Gary Gentile spent years working the legal system to gain the right to simply dive, and certainly not remove anything from, the site of the U.S.S. *Monitor*. He eventually won, and a small, distinguished cadre of divers including Hank Keatts and Billy Deans were permitted to dive the wreck. The legal lines were now more clearly drawn in regard to artifact recovery, but that did little to alleviate confusion in the ethical debate.

The controversy came to the forefront of my consciousness after years of dormancy while visiting "*Titanic* – the Artifact Exhibit" display in April of 2001 in Seattle. I overheard a worker explain to a tour group that all the exhibit's artifacts came from the wreck's debris field, and that nothing had been taken from the more intact portion of the wreck. To take artifacts from the intact portion of the wreck would not be proper because of the people who died on the *Titanic*, she continued. It was a curious drawing of the line, and one which begged the question just where exactly did the debris field end and the two large, separate "intact" pieces of the *Titanic* begin? It seemed logical that most of the *Titanic* victims were outside of the intact portions when they died, having either jumped in the water or been washed off the deck; did that make any difference in this rationale?

The distinction between "acceptable and unacceptable" areas to retrieve artifacts seemed to be a pretty big stretch, and it was at best a weak argument to rely on such an invented distinction to justify the taking of artifacts in an apparent attempt to avoid controversy. The fact that the artifacts were there for the world to see did not bother me personally, but

"SIMPLE ANSWERS AIN'T ALWAYS SO SIMPLE..."

that argument did. I had a lot more respect for the "don't take anything" argument, although I disagreed. Maybe this was simply the workers interpretation, but that was doubtful - it sounded practiced and had the ring of official spin.

The strict comparison to grave robbing was basically flawed in my opinion; it was not as if we were "backing our pickup trucks into the cemetery and hacking off pieces of tombstone," as Gary Gilligan used to say. The artifacts were not being hidden from the world either; trust me, anyone who possesses anything recovered from the *Andrea Doria* is champing at the bit to share the story and a look at the artifact with any poor sap that gets stuck listening. But the situation was not clear-cut.

John Moyer began to crate up his fantastic prize of the Gambone panels and truck it to various diving and nautical shows. It was an enviable artifact for any diver, anywhere in the world. And John had found it first. Or had he? Gary Gilligan casually let John know at one such show that he had discovered the artwork nine years earlier on a foray into the Winter Garden. In fact, while Craig and I were racing to bring my Promenade Deck window to the surface, Gary had struck out on his own. The Winter Garden had been his destination. Understandably, John Moyer was visibly skeptical. It was not uncommon for other divers to try and cut in on some of the attention; the only place I had seen egos comparable to wreck divers was on the flight deck of an aircraft carrier. Gary had obviously been waiting for this moment. He reached into his pocket, pulled out a small piece of colored stone and walked up to the Gambone panels on display. The small fragment of mosaic he had chiseled off nine years earlier fit a missing chip on John Moyer's prize perfectly. In fact, during the previous several years Gary Gilligan made several secretive attempts to retrieve the Gambone panels himself, once with a 300 pound lift bag. They were much too large, however, and over time John Moyer beat Gary to the prize. John Moyer was justifiably proud of raising, arguably, the most magnificent artifact since the *Top Cat* divers recovered the statue of Admiral Doria in 1964. But he had demonstrably not been the first to find art work.

SETTING THE HOOK

Billy Deans' fascination with the *Andrea Doria* did not end with the tragic death of his friend John Ormsby. He returned to the *Doria* on the *Wahoo* more than half a dozen times, bringing north from Florida with him the innovations that would make him famous in the technical diving community. Billy Deans' introduction of trimix to northeast wreck diving did not take off in popularity quickly, however. For years his charters were the only ones to utilize what were then considered exotic gas mixtures. In 1983 a first timer on the *Wahoo* had asked our group of four if any of us were diving mixed gas. The concept was so beyond our ken that it momentarily left us speechless. Craig made a quick recovery, however, and explained that no, none of us had eaten that afternoon at Taco Bell. That was the closest Craig, Gary, Don or I would come to breathing trimix until I started the course in 2000.

By the summer of 2000, about 40% of the divers on the *Wahoo* used trimix while diving the *Andrea Doria*, although virtually everyone utilized nitrox and/or oxygen for some form of accelerated decompression. Training was certainly an issue; it was a lengthy and expensive process to become trimix certified, particularly in the northeast where divers had to venture far off shore to reach the required depths. And the cost could be onerous, up to $150 to fill a set of double tanks with trimix, almost twice the cost of many day charters on a dive boat. Filling the same tanks with air would cost less than twenty dollars.

And trimix did not necessarily make diving safer. The only obvious commonality between the five divers to die in the 1998 and 1999 season (in addition to diving from the same charter boat) was that they were all diving on trimix. Was there a common cause after all? There probably was not any tangible, clearly identifiable link. The divers' experiences ranged from frighteningly little to over a thousand hours underwater. The accidents occurred at every phase of a dive: on the surface, halfway down the anchor line, inside the wreck, on top of the wreck, and on ascent. After reading the Coast Guard accident reports and newspaper accounts, the single common thread I could identify was distressingly ambiguous; each diver, in his own way, had pushed his personal limits too far. Was trimix, not directly, but through a subtle influence, responsible?

"SIMPLE ANSWERS AIN'T ALWAYS SO SIMPLE..."

The last to die, fifty-two year old Charles McGurr evidently passed out and drowned while he ascended up the anchor line in a heavy current. He had aborted his dive immediately after reaching the hull of the *Andrea Doria*. The coroner determined that the probable cause of death was heart failure. He had arteriosclerotic heart disease.

Christopher Murley never made it down the anchor line. He was incredibly inexperienced for the dive: only eighty-nine total dives, overweight at 350 pounds, and had a medical history that indicated obvious problems including hypertension and diabetes. He drowned in a panic at the surface while the crew from the *Seeker* vainly tried to assist him. He had been breathing trimix with a reduced oxygen content of 16%, barely enough to sustain the life of a person exerting themselves at the surface.

Vincent Napoliello tried to communicate some sort of distress to his buddy while in the silt out of the Dish Hole, but never could make it clear just exactly what was wrong. The pair exited the wreck, and Napoliello proceeded to swim away from his buddy, past the anchor line and out of sight. He was found minutes later dead on the surface. The isolator valve between his two main tanks was found closed; the tank he had been breathing from was empty. The probable cause of death was drowning after losing consciousness and exhausting the gas supply in the one open tank on his back. The medical examiner determined that he suffered from severe atherosclerosis of a coronary artery, which possibly contributed to a lack of blood flow, carbon dioxide buildup, and unconsciousness.

Richard Roost was an exceptionally experienced diver on his first expedition to the *Andrea Doria*. On his third dive to the wreck he did not surface. He was found lying in the interior of the Promenade Deck, not entangled in any manner, with no gas remaining in his tanks. He had earlier described the prospect of penetrating the Promenade Deck, with its many openings to light, as "easy." Roost made the dive alone and for an unknown reason (the autopsy found no obvious health problems) lost consciousness underwater, exhausted his gas supply, and drowned.

Craig Sicola had been to the *Andrea Doria* before, although he was not considered highly experienced inside the wreck. The accident investigation indicated that Sicola was extremely eager to recover china. He delayed his entry to the water during the mishap dive, possibly to separate from his buddies, and proceeded solo deep inside the wreck, through the First Class Dining Room to the kitchen area in search of china. He presumably got lost in the First Class Dining Room, eventually found his way out, but did not have enough gas to make it back to the anchor line. He lost control of his ascent, ran out of gas, surfaced unconscious and died soon after.

The cumulative result was five deaths in two seasons, and all mishap divers used trimix. Pushing a diver's personal limits had always been the primary cause of most diving accidents; it almost has to be by definition. But had these five divers overestimated their limits due to the prospect of diving without narcosis? Had the promise of technology's advantages tragically reduced their judgment, leading them to believe they were capable of dives that they would not have attempted on air?

The complexity, size, and weight of technical dive gear were no small factors to be discounted. It was frighteningly easy to breathe so heavily from exertion that a diver was literally fighting for control of their own body. The currents, challenging conditions, and pure excitement of a *Doria* dive tended to exaggerate this effect. The *Doria* was no place for a diver in ill health. It was a worse place for one lacking the experience to determine when a potentially disastrous situation was forming. There was no substitute for experience.

Had the angel of technology bought with her the demon of complacency? Was I heading east with the same disastrous reliance on technology as Captain Calamai during his race west decades earlier? I packed my bags and thought hard about the subtle traps that might await me, hidden on the wreck of the *Andrea Doria* and in my mind.

CHAPTER TWENTY ONE

Time to go

Summer, 2001

I kept in reasonably close touch with my high school buddy Don, AKA "Hook," over the years, and ran into Craig on an occasional visit back to the east coast, but it had been fifteen years since I last spoke with Gary Gilligan. All three had left the circle of *Andrea Doria* divers by 1992; Don and Craig had each accumulated more than fifty dives on the wreck, Gary one hundred. The three had earned their Coast Guard Captain's licenses and each ran dive or fishing charters part time, in Craig's case from the deck of the original thirty-two foot *Wahoo*, which he had purchased from Steve. They all left *Doria* diving for very different reasons.

Don Schnell's reason was medically driven when his left lung spontaneously collapsed after squeezing into a cramped space on a construction site. The ensuing examination at the hospital revealed that he had a not terribly uncommon malady of small air sacs collecting on the lining of his lungs, a condition called "blebs." Although not harmful in their current form, if one was to break free during a dive it would likely lead to an embolism and death. The fact that Don's first child had been born the year prior made it an easy decision and he quit diving.

Craig Steinmetz got caught up in a lawsuit after a diver lost his leg to one of the *Wahoo's* propellers in an anchoring accident. Craig was not directly involved, but the fact that he had been crewing on the boat that day made him one of the lawyer's targets. After years of going to court, Craig was ready to say; "It ain't no fun no more." He sold his business machine store, moved to Long Island's south shore, and became an independent trucker.

Gary Gilligan, still recovering financially from his bout with cancer and trying to save money to send his daughter to college (he was amicably divorced when we first met him), drifted away from the *Wahoo* on his own. But where Don and Craig stopped diving altogether, Gary kept it up, basing his diving primarily out of Connecticut. He lived on a small boat in Bridgeport where he had easy access to several wreck sites within a dozen miles of the marina.

It would have been worth money to see the look on Gary's face when he received a phone call from the Block Island Chief of Police toward the end of June, 2001. He didn't recognize my voice immediately, and he must have felt reasonably secure sitting on his small boat in Connecticut, being across state lines and all. Worse case he could cast off and try to "make a run for it" again. Don had told me during an earlier phone conversation that Gary was still actively diving, and it seemed like having him come along would be the perfect fit.

I let him in on the secret that it was really me, his long lost pal on the line (he probably would have preferred the Chief of Police), and we began to fill in the fifteen-year gap. Finally, I asked him if he wanted to dive the *Doria* again. After a two second pause, Gary responded, "Sure." You had to love the guy; he thought things through even less than I did.

Until that point, I had assumed the *Andrea Doria* dives would be made solo. Steve had initially expressed an interest to join me on the dives, but he was not much of an active diver these days, and as the time drew near I was reluctant to make him feel saddled by the commitment. A buddy could probably have been drummed up once out on the *Wahoo*, but it would undoubtedly be a stranger. My preference would be to go it alone if the only alternative was diving with someone I just met. Most of my dives in the past ten years had been solo, and it was uncomfortable trusting my life to a stranger whose background and intentions could never be made clear during a few hours on the deck of the *Wahoo*. Gary was another story. Despite not talking to him for fifteen years, we had shared so many experiences both underwater and above that we could pick up right where we left off. It would have been very surprising to me if Gary had changed much.

The dive charter that led to Gary's moped rampage on Block Island in 1982 had been his first experience on the *Wahoo*. We were on Block Island that August because the weather had precluded us from making it to the *Doria* on our first attempt, so instead we immersed ourselves in sun and Budweiser. Gary had a way of weaseling his way into another's confidence in a good sort of way. After Gary's run in with the Block Island police Steve had been a tad pissed off, and for the rest of the day Steve went on a tirade, telling Gary how screwed up he was, how no one on his boat had ever been arrested before, at least not while actually on a *Wahoo* charter. Steve had only allowed Gary to go on the *Doria* trip because Gary's local dive shop vouched for his experience, and Steve explained that once back in New York he was going to tell that dive store owner exactly how screwed up Gary really was. Two days later while drinking Budweiser on the aft deck of the *Wahoo*, Steve asked Gary to become a part of the crew. Craig, Don, and I presented him with our trademark "We don't drink beer, chase girls, or have fun - Ever" *Wahoo* T-shirt and Gary was officially part of the crew. What a weasel. Gary made the drive from Connecticut most weekends during the ensuing ten summers, none of us were arrested again, which made Steve happy, and we still managed to have fun. In the summers Craig, Don, Gary, and I would dive the New York wrecks. In the winters we would rent a Winnebago camper and drive to Florida, ostensibly to dive, but we brought along our trademark *Wahoo* T-shirts just in case. On reflection, the fact that none of us were arrested after Block Island was indeed significant.

I tried to keep the lighter memories front and center after encountering another speed bump in the pothole-strewn road back to the *Andrea Doria*. Steve cancelled the first *Doria* expedition two weeks before the planned departure date of July 2nd. The waning economy had created a down phase in the cyclical interest in extreme diving, only two people had signed up for the trip, and I was one of them. The *Wahoo* could not come close to breaking even. All my eggs were now in one basket and the weather had to hold for the second *Doria* trip - it was my only option. Steve assured me the second expedition was a definite go, but I thought if Gary went along as a paying passenger it couldn't hurt those odds. More importantly, having Gary along made me feel a hell of a lot more secure. It would certainly be interesting for the two of us to dive together again; I had updated my gear and dive style in every way imaginable to take advantage of the new technology, while Gary had been in a time warp back to 1982. Gary did take advantage of a wrist mounted decompression computer, which had become available in the late 1980s, and he used surfaced supplied oxygen for decompression, but other than that, he was the same old diver from the top of his leaky suit to the bottom of his well-worn fins.

The mismatch of gear, gas, and training was cause for some minor concern initially, but that evaporated after a single dive together. Seeing that I was stuck with the July vacation, the family joined me for a trip back to Montauk, New York to visit Don, his wife Chris, and their two boys. We got to see most of the old gang, even drink a beer or two, but more importantly I touched an east coast wreck for the first time in fifteen years.

Matt the Cat helped me carry my dive gear up the gangway of the passenger ferry for the day cruise to meet Gary in Block Island, where, true to form, he had spent a festive 4th of July weekend. After making the transfer of dive gear to Gary's boat, we left Block Island's New Harbor still enveloped in a shroud of fog to dive the wreck of the coal barge "*Montana.*" Gary certainly had not changed underwater, and neither had I. Our submerged high-fives and barely intelligible regulator grunts of "great dive" with hyped enthusiasm transported me back seventeen years. We would get along just fine.

But just to prove he was open to new ideas and equipment, Gary borrowed two sets of high capacity tanks and a wings style buoyancy compensator from his friend Joel Silverstein. After three warm up dives he was ready to go. But while I would be diving trimix, Gary would be breathing air. We planned on separating once on the *Andrea Doria* with Gary staying slightly shallower, and then hooking back up after our individual explorations. It felt much better knowing that at least part of the dives would be made with a trusted friend.

* * *

It was time to go. I sat in our living room, Sunday at noon, stared out the oversized windows that surveyed Dugualla Bay and Mount Baker, and finished a peanut butter and jelly sandwich. A small, sparrow darted from its perch on a telephone line along the street, raced toward me, and collided with the transparent window glass. The sparrow shook its small head, aimed its beak in the opposite direction and flew off. I felt a bizarre kinship. For ten months I had been banging my head against the barriers between the *Andrea Doria* and me. There was the training - forty hours of class work and forty-five dives in total: sixteen dives to the depth range to expect on the *Doria*, seven breathing air, nine utilizing trimix. My final pair of practice dives had been down a Canadian wall to 240 feet for thirty minutes on the bottom and a total dive time after decompression well in excess of two hours in the forty-eight degree water. I remembered the terror of almost dying during that dive with Ron Martin the previous October, the extravagant equipment purchases that had us living paycheck to paycheck, and my efforts to get the required time off from work. And then there was the first expedition's cancellation, another wrench in the works.

Most time at home not diving was spent watching the kids while Laurie traveled out of state as a consultant in the pharmaceutical industry. The distance that had grown between us reminded me of the Navy and the six-month deployments at sea. The pressure was intense, coordinating my job flying with responsibilities as a father and husband with what had become an irresistible - and, I began to consider uncomfortably, perhaps irrational - drive to complete my goal. Laurie was about at the end of her rope and I missed time spent with the family. I did not have to do this, there was no external pressure or obvious reward; middle age was ordinarily a

SETTING THE HOOK

comfortable, downright easy existence for me. The only glue that kept it together was my memory of how it had been seventeen years earlier.

Finally, today, the logistics were all worked out. The rental tanks for the dives would be waiting for me on the *Wahoo*; the rest of my dive gear had successfully arrived by Fedex at Don Schnell's house in Montauk. I bought a $200 oxygen analyzer at the last minute - Steve did not carry one aboard the *Wahoo* and my life depended on the oxygen content in each of the rental tanks. All that remained was to finish with final airline commitments and meet the *Wahoo* in Montauk before it departed Tuesday night.

I was scheduled to pilot a flight to Denver and then back to Seattle that same evening, and I planned to sack out for several hours at the Seattle airport Travel Lodge before catching the first flight east at 6:00 am. The rental car at New York's LaGuardia airport was arranged, all my last minute preparations complete. Nothing had been left to chance, nothing left to worry about except for that which was completely out of my control - the weather.

I hugged the kids and Laurie each longer than I ordinarily would when leaving for the same stint of time at work. Laurie quietly said, "I'm going to be really pissed if you get yourself killed." She did not smile or follow me out to the car to say goodbye.

I drove over the Deception Pass Bridge and craned my neck to look past my left shoulder at the cliff face that had started it all, where the twelve-year-old boy had separated from his family, fallen in the water and died. His body was never recovered; his family never experienced the relief of grim closure. I could not see the spot, the south wall of Deception Pass; it was too far behind me. I turned back around, punched the radio up loudly and felt my chest fill with the jazz of adrenaline and attitude. I was on my way, traveling back home figuratively and literally, back to New York, my youth, and my memories, back to how the future might have played out had I taken a different path.

I wanted answers, and it felt certain in my soul that they could be found if only I persevered. Where had that young man gone, the one who gave no thought to danger (or responsibility), whose carefree attitude permeated

TIME TO GO

everything he did? There was not a doubt in my mind that I would find him, buried beneath the comfortable clutter of children, bills, and commitments. My religious-like zeal, my gut wrenching need to merge with the past, was beyond reason or logic. If only I could dive the *Doria* again it would be all right; and as soon as that thought passed, I cringed at the first inkling of a different fear, a fear that perhaps things were not "all right" here and now, in the present. How had I missed that, and was it even a bit true?

My flight was delayed out of Denver that night. I skipped the hotel room after landing back in Seattle at 3:00 am and waited at the airport to start the final leg of my journey at 6:00 am. Twenty-eight hours after leaving for work and a restless hour or two of sleep on the plane, I arrived at Don's house in Montauk. The wind was blowing hard. I felt awful, completely out of my element - pretty shitty timing for a feeling like that.

Don and I drove out to a bar on the Montauk docks. He had a beer and I drank water, feeling the need to keep sharp, hell, to get sharp. We talked. Don had substituted his hunger for diving adventure in admirable fashion by volunteering as an emergency medical technician with the Montauk Fire Department. He was saving more lives now than he had while crewing on the *Wahoo* and had transformed his need for excitement into a strictly altruistic endeavor, well, except for the elaborate bar on the second floor of the Montauk firehouse.

We talked about the 1980s, our various adventures and antics, and the way it used to be with our group of friends. Without a doubt, Don missed the diving, but it sounded as though he had his fill of the "lighting our hair on fire" bullshit. I had to admit, in many ways so had I; hanging around Navy carrier pilots for ten years bridged that gap quite adequately. I couldn't say the same for the diving. Hook - I was the only one to still call him that - told me it never got better than in the early 1980s on the *Wahoo*. That was when we were all making our initial journeys to the *Doria*, when we surfaced with our first prizes, and felt the thrill undiluted by the later mass exodus of divers to the Everest of the depths. But I had nothing to compare it to, no string of experiences laced together by the gradual maturity of the wreck and the advance of the diving practices used to reach her decks.

The club the Atlantic Wreck Divers provided the *Wahoo* with an atmosphere similar to what our group of friends had offered during the second half of the 1980s. The sense of camaraderie and adventure was kept alive by this crew and the *Wahoo* was the king of the diving hill. The *Wahoo* continued to do some intense diving, including the *Republic*, the wreck found by Steve in 1981, which Don and the rest all experienced at 260 feet while breathing air. What old Hook-nose seemed to really miss, however, were the times before that, when much of what we were doing were either firsts, or damn close to firsts.

I figured he was trying to be nice, that he could sense my detachment from the scene, and was trying to make me feel as though I had not missed anything at all in the intervening years. But then again, maybe not, it was tough to tell. Several weeks prior I had spoken with Craig on the phone, and after explaining my plans to dive the *Doria* again he put it forward in his typically blunt manner with one simple question; "Haven't you had enough of that shit?"

I couldn't answer. Craig, Don, and Gary all knew for certain after seeing through their diving dreams to their individually determined ends, but not me.

Our conversation drifted away from diving, and then abruptly halted. The atmosphere became distant. I was heading out on the boat tomorrow and, for a change, Don was not. He was living here on the east coast, right where it was all happening, yet he was not going with us. I had traveled a long way and was nervous. We went back to Don's to catch some sleep. Tomorrow would be better.

When I woke up it was late and the sun streamed through the upstairs windows at Don's house. There were several hours to catch up and make ready for the *Wahoo's* arrival, so I brewed a pot of coffee, unpacked my boxes of dive gear, and grabbed a high protein energy bar that I threw in my bag at Whidbey. The soy derivative hunks of road apple looking putty were not bad tasting, and promised an instant energy shot in the arm when starting to drag. How trendy, I thought, sitting here in the old New York haunts eating this west coast granola-muncher's fluff.

Tired of waiting, I took a drive out to Montauk Point to gauge the wave action, and then went down to the docks to wait for the *Wahoo*. The pier was filled with tourists and spectators, viewing the catch brought in by the charter fishing boats and monitoring the progress of the few vessels that were only now returning to the harbor in the howling wind. It was a real-life watercolor of a east coast fishing town in summer: the commercial boats with their spools of nets and heavy steel dragger doors, the sport fishers filleting the locally caught stripers for their patrons, and parents and children wandering about in absent-minded bliss, apparently unaware of what lay offshore, the thrills, the history and challenge, the vital existence of a time gone by. I stood and simply watched in a detached state, a self-imposed barrier between me and the world. Thinking that the *Wahoo* must be running late, I went looking for Gary.

Figuring that it was a needle in a haystack (not really), I tried a local hangout that Gary had been known to frequent. During the summer of 1999, Gary had framed Don's new house while living on his boat, the "*Minnow Two*" at the docks in front of Liar's Saloon. This visit, Gary's beat-to-shit twenty-seven foot Silverton would not be there. I did not know what Gary was driving these days, and a quick scan of the parking lot offered no insight to Gary's presence. I forced myself to walk into Liar's Saloon, vowing that I wouldn't take a drink. Facing the bar and hanging T-shirts that sported the motto, "Montauk, a quaint drinking village with a fishing problem" sat my old partner in crime.

I quietly walked up to a barstool and sat down.

CHAPTER TWENTY TWO

"Exactly the same, only different…"

Summer, 2001, Montauk Point, New York

The sight of Gary sitting at a bar was a comfortingly familiar picture that could have been a snapshot from fifteen years earlier, at least from a distance. His blond mustache, California-looking tan, and brush haircut were all the same, but he was definitely older. At thirty-three Gary could pass for twenty-three; now at forty-eight, his leathery skin and eyes creased from what seemed a perpetual squint into the sun, made it difficult to give him more than a few years of free youth. But his manner and attitude were the same. He sat calmly at the bar drinking a Corona and smoking a Marlboro while making small talk with the bartender, who he called easily and often by first name. Having seen one another on Block Island just weeks earlier, we spoke only briefly while Gary finished his beer. After stubbing out the cigarette into an overflowing ash tray, we left for Don's house and to search out the *Wahoo*.

Standing at the base of the Montauk jetty on the harbor side, we watched the *Wahoo* round the channel entrance corner five minutes after our arrival. This was the same jetty where Don and I used to make night dives while in high school, catching lobster that we would occasionally sell to Don's Montauk neighbors for spending money. The *Wahoo* was still barely recognizable in the distance when we saddled up to return to Don's house to load my gear into Gary's truck and meet the boat.

SETTING THE HOOK

The pack up complete, we drove into the nearly empty asphalt parking lot next to the Coast Guard Station. The few cars present were overflowing with dive gear under the billowing clouds of gray sky: heavy steel double tanks, stage bottles, and bags hiding literally tons of miscellaneous dive gear. Tens of thousands of dollars of dive gear casually littered the asphalt waiting to be humped out to the boat. I took a quick survey of the gear-laden divers strung out like ants in two loose lines between the parking lot and the *Wahoo*: seven men including Gary and me, two women. It was a diverse group: some scruffy, bearded, one pony tail, a couple military length hair. Everyone seemed to be pleasantly relaxed. My eyes followed the trail out to the water until I found it; overhanging its extra length precariously beyond the short floating dock was the *Wahoo*.

The *Wahoo* was easily identifiable by her forward raked superstructure and sleek lines. The only boat in the New York area I knew with similar features was the *Seeker*. The *Wahoo* appeared deceivingly compact for a fifty-five foot vessel. As familiar as the image appeared, it was immediately apparent to me that something was fundamentally different. A single davit remained on her aft deck, the one where I had hung like a trophy shark going into Block Island was gone. The ladder to the helm topside had been moved from the stern to the bow, and now a wide row of wooden steps ran from the bridge to the *Wahoo's* forward topside "V." But there was something else, something of such significance that it should have been more obvious, but I couldn't put my finger on it. I slowly got out of Gary's pickup truck and walked with purpose to a bustling figure unloading the back of a Subaru Outback. Hank Garvin looked at me with a quizzical expression as I waved, which quickly turned to a wide grin of surprised recognition. He came forward.

"Peter! How the hell you doing?" Let out Hank as he gave my shoulders a rough hug.

"Great!" I replied, "How you been?" I shook loose from Hank's gruff hands and slapped him on the back. Hank Garvin looked only slightly older now at sixty-two, and his bulky frame was accentuated by the swooping

"EXACTLY THE SAME, ONLY DIFFERENT..."

bear hug. It did not appear his triple bypass of three years earlier was slowing him down in the least.

Hank Garvin had been going out on the *Wahoo* consistently since my first experiences on the boat. He started diving in the 1960s, but initially worked his way into the wreck scene at a slower pace than some of the rest of us. It was only after I left New York that he dove the *Andrea Doria*, but he sure made up for lost time. Don let me know over the years that Hank was into some seriously deep wreck penetrations. As the rest of us let up in our diving activity, Hank Garvin's interest accelerated. Hank was an insurance agent and now, after Steve and Janet, the third working captain on the *Wahoo*. He started diving mixed gas fairly early in the game, and was one of the most experienced *Andrea Doria* divers around. Hank lived by his routine and rarely missed a diving trip; in fact, he conducted much of his summer business from the boat on his cell phone. He would be the only person in addition to Steve, Janet, and Gary that I recognized from the old days on the *Wahoo*.

I shifted my gaze over Hank's shoulder to a figure lumbering down the dock. The deliberately swaggering bruiser's gait of Steve Bielenda was easy to identify, and I'd be damned if he didn't look the same. Sure, his hair was completely white now, but his face and features appeared to be etched more by his character than the years. I never saw Steve when he was truly young, but I doubted that his round, full Polish face would ever look truly old. I walked up to him with a suppressed grin and held out my hand.

"How are ya, Peter?" He said in straight-faced pure Brooklyn. Steve grasped my hand firmly in his paw.

"Good Steve, real good. How are you?" Steve, unlike Hank Garvin, was definitely not the hugging type. Conscious restraint, the need to always keep something in reserve, was one of his hallmarks.

"What the hell is this?" I flicked my right hand by Steve's head at an earring. I looked closer and saw the small silver ring was actually a shackle. Perhaps because of brand loyalty to the *Wahoo* years prior, our small group of friends was the only cadre I knew that could flick Steve shit with relative impunity.

SETTING THE HOOK

"Well you know how it is," I supposed this was the closest Steve ever got to bashful. It was not very close. After first meeting Steve as a college kid, I briefly had an earring, a small gold stud. Earrings were not universally accepted, never mind fashionable, in 1982, I caught a load of shit from a lot of people on the *Wahoo* for having one, and Steve had been the ringleader in my good natured, but unrelenting persecution. I kept the earring long enough just to show that I was not succumbing to peer pressure, which, of course, was a lie, and then got rid of it - my, how the worm had turned. It was apparent that Steve remembered this small fact of the past, and I easily slipped into a role left decades earlier and reveled in the silence of knowing that I had gotten him good with this one.

Steve and I had been in contact by phone or email consistently for the last few months, and we didn't engage in idle talk for long. There was still much for me to do before departure, and I turned and walked quickly toward the *Wahoo*. It was only when much closer, barely a hundred feet away, that the *Wahoo's* difficult to pinpoint change became clear - she was plain worn down. The blue stripes along her side, even the large letters of "*Wahoo*" itself, were faded, and there were gaps in spots, weathered and worn from the years. The corrosive overrun from her drains permanently discolored her hull in narrow strips at the stern, and one broad swath of rusty brown ran halfway up from the waterline at the bow. She was a skeleton of my memory's image, the shape and form unchanged, all permanently connecting her character and stories; yet completely different in such a basic way that it was unsettling.

As Craig liked to say, "Exactly the same, only different." I wondered what her engines sounded like.

Steve directed me to the stowed rental tanks he was providing for a "small" fee - always the businessman - and I got to work analyzing the gases and adjusting the double tank bands to fit my gear. Halfway complete with what was turning into a major project, Janet emerged from topside where she had been taking a nap. Janet had just completed the eight hour trip from the Captree Boat Basin to Montauk and was now about to make the ten hour run out to the *Doria*, during which time she could expect

"EXACTLY THE SAME, ONLY DIFFERENT..."

only an occasional break at the helm from Steve and Hank. The two men were well rested, having driven to meet the *Wahoo* in Montauk, but the *Wahoo* was clearly Janet's baby now. She ran the show, and the other two captains were essentially along for the ride.

Janet, like Steve, did not appear to have changed greatly. If anything she was slightly larger, but her hair was now cropped short, creating an almost unbalanced impression compared to my recollections. When I went up for a hug, her exaggerated, almost coy expression made it clear she was the same. She was a shy, little girl at heart.

I went back to setting up my equipment, not wanting to be fiddling with wrenches once the boat deck started pitching. It ended up taking longer than expected, and by the time my gear was adjusted and assembled it was almost 6:00 pm. The last minute rush was frustrating; after all my careful planning, damn the expense, and here I was racing the clock before the *Wahoo* pulled out. This was normal with any rental gear, adjustments were almost always required, but I had counted on getting a break from the weather, and had expected calm seas that would allow handling the tedious but important job at my leisure on the long run out to the wreck. All the others except for Gary had headed out for a relaxed dinner well prior to the 7:00 pm departure, and finally, with just forty-five minutes remaining before cast off, we raced over to Westlake restaurant to catch up with the group.

Don and his family joined us, and we sat at a table next to Steve, Janet, and several of the paying charter members. I flagged down the waiter, eager to finish dinner and make my way back to the *Wahoo* and the unknown. It was all so different from the memories. We used to leave much later, after dark, from where the *Wahoo* would be docked at the town pier. We would make the short walk to Saliver's bar with its walls lined with trophy shark heads and have a last dinner ashore. It was a much changed ambiance to the outdoor tables at Westlake, well inside the harbor in the light of day. No transient boats were allowed at the town dock now - the rotten wood was falling apart, and probably would have been condemned already if there had been anyplace else to park the small commercial fishing fleet.

SETTING THE HOOK

I tried to push the ordering along so that I could return to the *Wahoo* and finish preparing. My last chore before heading out to dinner had been analyzing the thirteen rental tanks to determine their exact gas compositions. Now armed with that data, I needed to run both planned and emergency contingency dive profiles on my laptop computer. These dozens of lines of abbreviated annotations for times, depths, and gas mixtures would then need to be transposed in pencil onto my wrist slate, hopefully before we were out in the waves of the open ocean.

I settled quickly on a chicken sandwich, a suitably bland fare to try and balance the whirlwind of my mood. Steve's loud voice was clearly audible at the next table.

"Look at him, Janet," Steve's Brooklyn accent was thick, almost exaggerated, "He's not a kid anymore."

"I don't see much gray hair, but you're right." Janet answered while matching Steve's overdone side-glance for my benefit.

I looked over, feeling out of place. In the past two hours, I had automatically assumed my old role on the boat as the young, brash college kid, trying to make up for life experience and knowledge with bluster and energy. Maybe that was why everything seemed uncomfortable; it was not me anymore, but I felt compelled to fill the mold, to latch onto any role that would keep me from floating ungrounded during our impending adventure. The need to explain that the intervening seventeen years had not passed me by, that I had grown fundamentally more than they could imagine, was suddenly important. I wanted to remind them of my experiences, the Navy carrier missions, of my time at war, my marriage of nine years, and my two wonderful kids that meant more to me than anything in the world. Steve and Janet knew all of that anyway - why did I feel compelled to tell them again?

"You're just too far away, there's plenty of gray, Janet." I yelled back across two tables. There was no audible retort. We quickly finished our meals.

I jumped out as Gary's truck came to a halt and walked quickly back aboard the *Wahoo* and booted up the laptop. Halfway through the slate

"EXACTLY THE SAME, ONLY DIFFERENT..."

writing, the *Wahoo's* twin diesels rumbled loudly to life within seconds of one another. I paused, looked up, and took a deep breath; it was the only aspect of the *Wahoo* that was identical to my memory, that distinctive, guttural roar to life of her engines, the subtle waft of diesel fumes. Instantly, for only a moment, I was transported back, to the countless times awakened by the comforting wave of engine noise after a late night of carousing. We would stagger up out of the Doghouse, help untie the dock lines, and then trudge back to our racks for another two hours of sleep before we reached the wreck site and it was time to set the hook.

The soul of the *Wahoo*, her diesel engines, was the same.

I finished crunching the dive profiles, closed the laptop computer, and listened while the diesels warmed up. The only other time I had touched a keyboard in the vicinity of the *Wahoo* in the pre-personal computer days of 1983 was hefting that old, corroded piece of shit typewriter out of the Dish Hole with Craig Steinmetz. I looked over at the youngest of the charter on board, Heather and Dave. They were a friendly couple who ran a charter dive boat out of Massachusetts. Both were 26, about four years older than me the last time I had been on the *Wahoo* preparing for a *Doria* trip. I wondered if they had ever used a typewriter before; I wondered if they had ever seen one as adults.

Don ducked inside the cabin, walked up to me and reached across the laptop to shake my hand.

"Good luck Pete, have fun." I could not tell if the edge to his eyes was concern or envy, or both.

"Thanks, 'Dan.' I'll call you on our way back in." On a cell phone, of course, I thought, something else unheard of in the old days. I was Don's best man at his wedding. During my traditional toast to Don, with me in Navy uniform and pilot wings, I kept calling him "Dan" just to throw the hundred guests off track. Don's mom's friends would hound her for years after: "That best man must have really been drunk, he kept calling your son 'Dan' at his wedding." Although I had been drinking to be sure, I knew exactly what I was doing. I was the only one to call him Dan, and usually only on occasions of unique strangeness.

SETTING THE HOOK

I just finished zipping the laptop case closed when the lines were cast off and the *Wahoo* pushed forward. It was impossible to ignore the darkening sky or the whistling wind in the distance, and while it might be calm here in the harbor, that clearly would not last. I made my way to the aft deck rail. The *Wahoo* rumbled out between Montauk's twin jetties and turned into the wind and building seas. Long Island's South Fork would still protect us from the southerly wind and ocean waves for a few more miles, but when we rounded Montauk Point by the lighthouse we could expect to bear the full brunt of the ocean's energy. I wasn't looking forward to the ride.

For ten months I had sweated the details and worried about the individual aspects of my grand plan - finishing each phase of training, getting on the *Wahoo* charter, buying and putting my dive gear in order - and now here I was, steaming back in time to the wreck of the *Andrea Doria*. I was truly obsessed, perhaps with the type of "mission oriented" mindset that I was so critical of in other sport divers. But mine was not just a diving journey. I traveled the road more in memory than in action, and it wasn't easy, not certain even now that I was ready to be confronted with what I had actually become. I was far more nervous than ever before in anticipation of a *Doria* dive. But I had to know if it were possible to re-capture that unfettered and clear-cut happiness of youthful irresponsibility. This penetration by my soul of the wreck of the past was in search of the most elusive of artifacts - the wide-eyed simple freedom of young adventure. But the closer I got to my goal the greater became the tug of my family, as if my soul had secured a penetration line firmly to the present, and was only with increasing resistance willing to pay out the distance yet to explore.

The Montauk lighthouse came into view off the *Wahoo's* starboard bow. The lighthouse overlooked the magnificence of the Atlantic Ocean to the south and Long Island Sound to the North. Seventeen miles east was Block Island. Perched high on a hill with its beacon unobstructed for 360 degrees, the lighthouse stood sentinel for generations of mariners as they struck out for the ocean's mystery. At the lighthouse base was a somber

"EXACTLY THE SAME, ONLY DIFFERENT..."

black memorial listing the names of all the fishermen who had died after leaving Montauk Harbor, most of them killed by the turbulent sea. The list was not short.

The sun was in its final phase when we rounded the corner of Long Island's South Fork, jogged heading to the south of east, and felt the full power of the Atlantic. The spray reached up over the *Wahoo's* gunnels as the ocean swell tossed her with haphazard indifference. The subtle sickness in my stomach was not from the wave action, at least not yet; it was from the realization that these conditions, here, only two miles off shore, were not dive-able. The *Wahoo's* props lifted slightly out of the water after a particularly large wave and cavitated loudly in the dusk air.

We had almost a hundred miles of open-ocean yet to traverse. I knew in my gut we would not make it, at least not tonight. I balanced behind the *Wahoo's* superstructure, holding onto the top of the doorframe where I was shielded from the spray, but still in the fresh air where the diesel fumes from the engines were blown well aft.

Gary went below to the doghouse to sleep. Two of the charter members braved the swell and spray to sit in the fresh breeze of the bow. Everyone else was inside. I found myself alone on the *Wahoo's* aft deck, alone and sober, the sun almost down, the sky growing more ominously gray and foreboding by the second. This was not at all as I remembered things, but why the hell should it be, even if my memory had not deluded me into only remembering those days of pristine conditions. I was still not used to the waves, the open Atlantic, the growth of the *Wahoo* from a new, sparkling hull to an aged workhorse.

I took a last look at the sun before it departed the sky. It was not where it was supposed to be. I doubt I would ever have realized our change in course intuitively by the setting sun seventeen years ago, but now it was obvious. Ten years in the Navy and sixteen years of navigation and flying were worth something out here, I supposed. But what was obvious to me was not good news. We were undoubtedly heading for Block Island.

Gary Gilligan and the author in front of Ballards in 2001. From the author's collection.

Steve Bielenda on the *Wahoo* in 2001. From the author's collection.

"EXACTLY THE SAME, ONLY DIFFERENT..."

THE *WAHOO* PIER SIDE BLOCK ISLAND IN 2001. FROM THE AUTHOR'S COLLECTION.

CHAPTER TWENTY THREE

Six Pack Rock

Summer, 2001, Block Island, Rhode Island

Ballards is timeless. Despite the building having burned down in 1986, the scene before me was identical to the one branded in my memory. Ballards had been rebuilt to its original design after the fire, but it was more than that. Sitting at the island bar with the band loud at our backs, the diners sprawled across the expansive acre of tables, and a cool sea breeze blowing through open windows, it was all the same. But to be fair, I had been here once since the demise of the original structure.

I scanned the room, searching the dozens of pieces of memorabilia that decorated the walls with the names of fishing vessels and parties of revelers that had left their mark. There it was, in a different spot from where I had put it ten years ago, a plaque showcasing a photo of the *Wahoo*. Stuck to the very bottom of the plaque was a small colorful emblem. If you got close you could detect a military aura about it, but to the uninitiated it was meaningless.

The last time I had been to Ballards - the only time since 1985 - was 4th of July weekend, 1991, while on leave from the Navy after flying combat in Operation Desert Storm. I had placed the A-6 Intruder squadron sticker showing a Griffin and Sword on the plaque at the end of the weekend. Looking at the sticker and plaque, now divorced from both of these defining experiences of my life, was surreal. It was the only physical connection

SETTING THE HOOK

I could think of between the two parallel adventures of my growing up, the *Wahoo* and the Navy - a damn sticker. At least it was at Ballards.

"What do you want, Pete?" Gary asked. His face carried its typical unexpressive intensity.

I swiveled in the low-backed chair, pulled closer to the bar and yelled to be heard above the Irish jig. It was a different band than the one that used to play at Ballards, but still, it sounded hauntingly familiar. Maybe it was just the resonance of the lively music bouncing off the liquor soaked walls, or maybe all Irish music sounds the same to me. It was crowded, despite being an otherwise non-descript Tuesday night in the summer.

"What do they have on tap, is that Bass?" I answered, conscious of my hair sticking in all directions from the wind we had just left. The bartender nodded yes. "Sure, I'll take a Bass." Gary yelled to the bartender, and I watched as he quickly made a Black Russian and poured a Bass ale.

Eleven pm - this was not exactly where I expected to be right now. We should have been thirty-five miles off shore, heading southeast. We were facing southeast at our perch at the bar, and that would have to be close enough, at least for tonight. Hank Garvin, Janet, and Steve walked in and sat down next to us. Trailing were several others from the charter. Steve and Janet got locked in some conversation, unquestionably about the boat, but oblivious to the rest of us in the confusion of loud music.

"So Hank," I almost shouted, "Is Gimbel's Hole really closed up?" Gary had spoken with several friends who had dove the *Doria* during the 2000 season, and they indicated that the wreck was deteriorating so badly that Gimbel's Hole was now impassable.

Hank answered in his luggy New York accent, "Completely closed? I don't know about that. It sure as hell is falling in on itself, but I think you can still get through it."

Overall, it sounded like there was not much access remaining to the original Dish Hole. Prior to the removal of the Foyer Deck doors there was only that tiny cut out from the failed 1973 commercial salvage attempt. That was before all this shit started, before charter boats began running regularly out to the *Doria*. The *Andrea Doria* had allowed the passage to remain open for twenty years. Why close up now? A chapter in her history closed with that hole, it had to, didn't it?

"Is there any other way to get down to the Foyer Deck, over by the dining room?" I directed the question again to Hank.

"Yeah, most of the higher decks have fallen off. You can pretty much find a spot at the Upper Deck on her side and go right in, but it's tough to tell exactly where you are. I went into the kitchen a few years ago. That's just past the dining room through the passage where all the dishes were, a pretty deep penetration. When I was getting ready to leave, I looked around and saw a big hole in the side! I said to myself, the hell with this," Hank hitch hiked his thumb behind him, "I'm not swimming back through all that crap to go up through Gimbel's Hole. I ended up out on the *Doria's* deck, on her side. Swam right up and there was the anchor line."

I had to let that settle for a while. The wreck of the *Andrea Doria* had been defined for me by Gimbel's Hole. That was what started us all going out there. Out of my thirteen dives on the shipwreck, ten of them had started by dropping down through Gimbel's Hole.

"So where is the china coming from now, if you can't easily get to the old spot?" I asked.

Gary cut in. "A lot of the guys are supposedly going down to the sand at the base of the Foyer Deck, in the washout. The hulls broke open enough that some china's spilled out. Mostly broken, I think."

"Shit, 250 feet for broken china." I said, more to myself than Hank or Gary.

"Try 260 Pete." Hank chimed in, slowly lifting his drink. I noticed it was a Pearl Harbor - vodka, Midori, and pineapple juice. Pearl Harbors used to be a staple for us at Ballards whenever we wanted to get "bombed," in other words, most of the time.

Hank continued. "Last year when we tried to grapnel into the wreck the hook kept skipping off the top. The Boat Deck and most of the Promenade Deck have slid down into the sand. There's not much left to catch." Hank took a sip. "Pete Manchee and I went down to set the hook, and the thing was caught in a net that had fallen off into the sand. What a screw job."

Pete Manchee was crewing on this trip. Gary knew him, but I did not. I remembered he was mentioned in Gary Gentile's "The Technical Diving Handbook" as one of the guys that he dived the German battleship *Ostfriesland* with in 1990. The *Ostfriesland* was in 380 feet of water off Virginia.

"So I'm trying to pull the hook out of the net," Hank went on, "and Pete is pulling in the line against the current to keep it slack. I look at my depth gauge, 260 feet. Beats the hell out of me where the wash out ends now." The digging action of the hull into the sand had been consistent and patient, as if the *Doria* were trying to bury herself.

"Finally, I get the grapnel loose, and Pete swims it up to the rail. He's got one hand on the bow rail, one hand on the grapnel, and his arms are like this." Hank spread his arms out into a wide cross. "That was the end to the slack in the line. Five more feet and we would have been hooked in. We ended up saying screw this, went up, and let someone else try. We finally get hooked in, but all three of my *Doria* dives last season were working that damn grapnel."

I thought about it. Two hundred and sixty feet down to scrounge around for broken china. In many respects I maintained my references to depth with the mind-set of breathing air. Two hundred and sixty feet was too damn deep to be doing that on air as far as I was concerned, but with trimix maybe it wasn't that bad. On the other hand, decompression was decompression; a lot could still go wrong.

We slowly finished our drinks, each of us sticking to two. Tomorrow, when the wind subsided and we took another shot at it, nobody wanted to be hung over. Still on west coast time, it was 1:00 am by the time I decided to get some sleep and made my way down to the doghouse, to the tiny open racks that remained. The layout of the doghouse had been completely redone, all the spacing and partitioning of sleeping areas was different. I never considered racing down earlier to claim a bunk, but the others obviously had.

Gary and I were left with two very narrow, low bunks up against the bulkhead where the diesel engines roared while underway. The bulkhead still radiated heat despite the shutting down of the engines hours earlier. I could barely fit my arms in. I found out later that these two choice bunks were affectionately called the "spice racks." Affectionately by those who did not have to sleep in them.

This never would have bothered me in the old days. There was only marginally less room than our racks on the aircraft carrier, where I had spent six months at a shot. These days I spent a lot of days on the road, crisscrossing the country with United, a lot of time in strange beds. We almost always stayed at very nice hotels. I got up after fifteen minutes and trudged up to the bow. You could get beat to shit up at the bow berthing while underway, but here at dock it didn't make any difference. I found a wide-open, spacious rack with a fan and tons of legroom and after an hour finally fell asleep.

Brilliant streams of sunlight woke me the next morning, and I lay in the bunk and listened to the dozens of flag banners lining the beach in front of Ballards flap in the wind. There was no need to climb the ladder from the bow berthing to see if the weather had improved. The sun was only a tonic for the emotions; the wind was what dictated the state of the seas. I looked at the readout on my digital watch; 7:30 am, the 25th of July. I pulled my way up the ladder to join the world.

Steve and a group had just set out for breakfast in town and Gary had just returned from breakfast solo. A byproduct of early rising in the construction trade, Gary was fond of saying that he was useless without breakfast. Or with breakfast, I would be quick to point out. While Craig, Don, and I would be nursing our hangovers with a cup of coffee on the *Wahoo*, Gary would be heading off, often by himself, to get something to eat. Gary sat on a dock pylon smoking a Marlboro.

"Morn'n Gary." I mumbled. "You want to sit around and watch me eat?"

"No, I'm going to relax here for a few. A bunch just left for the café up the street, go catch 'em." Gary answered.

SETTING THE HOOK

"Right." I turned to go.

I stumbled off the *Wahoo* to wander down the dock to the community washroom to clean up before town. The sky was infinite blue, not a cloud to be seen, no reference other than the sun. On the water, behind the protection of the island and the jetties, the wind could not reach us. Without clouds to fill in the edges of my senses I needed to look at the flags on the beach, or the one on the roof of the National Hotel, to guess at the wind speed. The flags beat in a flurry with each gust, but stayed horizontal throughout. We were not going anywhere today.

With reluctant acceptance, I tried to step lively up the street, already crowded with tourists and cars. It really was beautiful, the sun completely unobstructed, Old Harbor's asphalt streets warmed in the calm created by the blocking action of the rustic buildings, each unique style adding character to the scene. Half the diners in the tiny café were from the *Wahoo*. I sat down and tried to make light conversation, but there was a tiredness in my bones I could not shake. I ate my omelet mostly in silence, frequently sipping coffee, maintaining a weary edginess, listening hopefully for silence from outside. The distant whip of wind offered no respite.

Steve spoke up. "Peter, we're going down to the museum in a few, coming?"

"Sure," I replied, and tried to think of anything to say to keep the conversation going. I came up blank.

We came to the small Block Island Museum - originally a 19th century house - paid our modest entry fee, and walked the rooms in silence. Dutch Captain Adrian Block sailed by the island in 1661, but never set foot on what would become his namesake. Block Island's history revolved primarily, as would be expected, around fishing. When first discovered by Europeans (there were an indigenous peoples already living there), there was no natural harbor. Around 1676 a breach was dug on the island's northwest side to open up a great salt pond to the ocean. The breach was maintained with the help of shovels and sweat against the sea's constant efforts to reclaim her rights until 1895, when the first permanent passage was built. Oddly enough, this first protected anchorage was named "New

Harbor." "Old Harbor," on the other side of the island where the *Wahoo* was docked, had no natural advantage of topography other than being in the lee of the prevailing winds. Two huge jetties had been constructed so that Old Harbor could be used in any weather.

The ferries and larger vessels came into Old Harbor. It seemed that nature would allow continued access into New Harbor, at least with advanced dredging techniques, but it was the completely manmade Old Harbor where a town had developed and commerce existed. The combination of sea and time were not to be trifled with, and the line between nature and fate was blurry at best.

We killed as much time as possible in that museum. Other than a brief detour into the local dive shop - where wreck diving books by Hank Keatts and Gary Gentile were prominently displayed - that was the extent of the formal events of the day. The activity de jour seemed to be to hang around the boat and look at the dive gear. I certainly did my share.

After several hours of meaningless wandering, frequently glancing up at the flags on their poles, Steve surprised Gary and me by announcing that he would entertain staying out an extra day if it looked like the winds would ease. That would mean the cancellation of a planned *Wahoo* charter from New York on Saturday, definitely not Steve's modus operandi. It was understood by all aboard that there were no refunds, no rain checks for a *Doria* trip; you paid your dime and you took your chances. Steve's offer was a novelty, a break from a long tradition of sticking to the original deal, to make as much money as possible in the summer so he could justify the *Wahoo's* existence for the long winter months when she sat dormant. I wondered whether he too, in his own way, was attempting to wait for the past, for the "glory" days of *Doria* diving, carefully bracketing his quest with the reality of the present. I could tell Steve missed having Gary and me aboard.

I felt momentary elation that it was still possible; there was time if the wind died down by early morning to make it out to the *Doria*, at least for one day. I sat down. It only took an hour of staring at the rigid flags, unyielding in the slightest, to begin to drag me back to reality.

I knew then that we would not make it.

SETTING THE HOOK

A completely bland disappointment enveloped me, bereft of emotion, a nothingness that settled deeply in my being. I would not see the *Andrea Doria* I had once known, not tomorrow, perhaps not ever. I would never decipher the wreck's message to me, that rhythm spelled out so many years ago in the reversed dripping of bunker oil to the glassy flat surface. There was nothing to say or even to think about it, it just was. The words of Matt the Cat, the Bronx bus-fixing philosopher echoed in my mind, "Peter, sometimes things just be's that way." I sat and shook my head.

Gary brought me out of my slump, shoulders hunched in the sun on the *Wahoo's* aft deck, with a simple question: "Hey Pete, wanna go swimming?"

"Yeah, sure." I said, mustering my energy, not at all certain that I wanted to go anywhere near the ocean. "Let me change."

Five minutes later we were in the chilled, refreshing Atlantic, bobbing under the brilliant sun in the building waves I had been cursing under my breath all day. It felt good to let the blood move through my body, to feel the cold wash over my skin. In the old days, we would occasionally swim the short distance off this same beach to a rock that was exposed at low tide. After scraping our legs until they bled on the barnacles, we would sit hunched into the waves and drink from a six-pack of Budweiser that we dragged out with us.

Sitting on "Six Pack Rock" was the closest, I believe, our small group came to being truly contemplative. We would sit with legs dangling in the cool water, the blinding sun warming us from without, the beer from within, and simply look around. On our modest throne we were kings of our destiny, the world an infinite possibility and the future unreal. Who the hell cared, everything we could want was right here, in front of us, waiting to be experienced, patient with our delays, full of promise and mystery. When it was time to go back to the beach, our beer gone, our spirit bursting beyond the bounds of spectating any longer, I would stand up precariously on the rock, balancing into the breaking waves, the blood running down my legs, and thunder into the wind, "Is there no one to challenge me?"

Of course it was a joke. The words from a "Superman" movie sequel lent just enough mock gravity to the situation to allow us to dive into the

SIX PACK ROCK

water and race back to Ballards, the loser damned with buying the first round. Still, the power I felt in my voice was real. I was up for anything.

Gary and I swam for almost an hour, treading water, slowly working our way up and down the beach, talking very little. We looked out at Six Pack Rock. I never considered going out there, and I doubt Gary did either. My poisoned spirit gradually seeped out, allowing room for a rejuvenated sense of energy, a fullness of life. Dunking my head into the cool water was more than refreshing; it was the culmination of a circle, I had been here before. In spite of its eternal indifference, perhaps because of it, the ocean was slowly bringing me to terms. I walked out of the water not exactly with a spring in my step, but a balanced sense of direction. We washed off the salt with a hose that had been left coiled on the dock in the warm sun, took one last look at the flags in the wind, and walked over to Ballards.

Over the past twenty years, from the time of Steve's first charter to the *Andrea Doria*, only four trips were cancelled due to weather. I was on the first blown out expedition in 1982, and now I was on the last. Since the start of the exodus of divers to the *Andrea Doria* in 1981, 2001 was the first year that the *Wahoo* did not make it out to the shipwreck at least once; exactly twenty years to the day - absolute coincidence.

Forty-five years to the second after the *Stockholm* hit the *Andrea Doria*, Gary and I sat in Ballards, telling old stories, reliving memories and experiences. We did not notice the event; I only realized this much later. The next morning, possibly at the precise moment that the *Andrea Doria* left the surface forever decades earlier (I don't know for certain, I did not think to check my watch), the *Wahoo* left Block Island in what were worse seas than two days before. The sun was completely hidden behind the rapidly darkening sky. A blowing mist filled the sea air, a curiously incongruous fog coexisting with the driving wind. We headed back to Montauk.

I had set the hook in a quest for my past, traveled as far as reality would allow, and now it was time to free the hook and head home, back to my life. I returned to Washington State, to Laurie, Emily, Jared. Soon after arriving, I was blessed with the comical vision of watching three-year-old Jared attempt to march around the living room with my dive fins on his feet, and

SETTING THE HOOK

my old Navy flight helmet on his head. He never quite got the hang of it, but I like to think I learned something by watching him as he stumbled into his future. I was clearly shackled by my children, but to be free of them would be a horrendously worse nightmare than I could ever imagine coming from the *Andrea Doria*. Perhaps the *Doria's* message to me was not lost. Maybe more can be learned from life by listening than by prying.

The prosperity and technology we Americans enjoy so greatly feed the illusion that we can re-live the road not taken to experience what might have been. With persistence I knew I could return to the *Andrea Doria* herself: the weather, regardless of how influenced by fate, would change. But so do all things. The shipwreck is no longer as I remembered her, and neither am I. The spirit of the *Andrea Doria* was not calling me, what I heard was the subtle yet distinct summoning of my own soul. If memory is a separate reality then it is just as unreachable as another universe - it can only be observed through the telescope of the mind.

The unrelenting winds of time push us forward; away from the fog of the past, but if we are not careful, also away from our goal. What is important is the journey, the required solitary exploration; our plans and dreams rarely mimic the reality of outcome. I realize now that it is more difficult to let go than to doggedly pursue a goal, particularly an unattainable goal. Pursuit merely requires effort; release requires contemplation and acknowledgement.

Before my Odyssey to the *Andrea Doria* I would look blankly beyond and through that upstairs curio cabinet of Italia china, not really seeing or understanding. During my nearly year of preparation, I fixated on each dish and cup with trepidation and fearful promise, hopelessly losing the forest from the trees. Now I gaze at the display with contented acceptance for where I have been and where I might travel.

In life - as in diving - there is no substitute for experience. I let go.

A LONE DIVER APPROACHES THE ANCHOR LINE HOOKED INTO THE *ANDREA DORIA*'S AFT CARGO BOOM AT THE END OF HIS DIVE. PHOTO COURTESY OF BRADLEY SHEARD (1989).

EPILOGUE

The Third Decade

> "...And though you come out of each grueling bout,
> All broken and beaten and scarred,
> Just have one more try — it's dead easy to die,
> It's the keeping-on-living that's hard."

Final line from the *"The Quitter"* by Robert W. Service

Spring of 2011

I didn't dive a great deal for several years after returning to Whidbey Island in 2001. The big distraction was initially the fallout from the September 11th terrorist attacks, which, granted, was a fairly huge distraction, particularly to a United Airlines 757/767 airline pilot like me. But my waning interest in diving stayed after America's "new normal" arrived, which ultimately didn't seem much different from the "old normal" anyway. My dozen or so annual dives were mostly a routine of habit with long time spear fishing cronies, which had always been more social event than serious diving. I finally realized just how much ego had been woven into my wreck diving memories, which did nothing to distort the facts of the past, but definitely had misdirected a balanced understanding of the now. It was clear that the time and money commitments of deep trimix were not for me; I had no desire to dive on the edge anymore, in fact, I was no longer sure where to find "the edge," and did not particularly care.

The near mass market appeal of trimix had eroded the primary barrier to making deep dives, the narcosis, and in some respects deep diving had become downright pedestrian. Where it used to be enough to simply tell a fellow technical diver that one had "dove the *Andrea Doria*" to gain credibility, now it had to be explained that the dives were made breathing air deep inside the wreck, and then go on to describe why this was fundamentally different from the mixed gas equipment-laden technical diving of today. Diving credibility used to be important to me, but today, again, I don't particularly care. It seemed that large segments of new tech divers were leapfrogging the old prerequisite of building practical experience incrementally with expensive advanced training and apparently getting away with it, at least in the short term. A non-diver friend previewed this book and related my sentiments nicely to his mountain climbing experience:

"I try to continue what used to be easy and carefree and novel, only to realize I can't do it anymore, or more insidiously I don't care enough to take the risks, make the investment."

It seemed appropriate to close out my *Andrea Doria* memories with a new and different mountain climbing analogy so as to not leave the "Mount Everest of wreck diving" phrase hanging in an endless decompression.

To be fair, there was one other contributing factor to my retreat from diving, a big one; at forty-three, I was diagnosed with young-onset Parkinson's Disease, an incurable debilitating neurological disorder which tends to make all physical activity progressively more difficult. No one knows what causes Parkinson's, although some of the theories lead me to believe that it either came from exposure to neural toxins during the first Gulf War or maybe from high oxygen partial pressures or repeated instances of silent bubbles from diving deep, or maybe a combination of the two, or maybe something not considered; it doesn't matter, I own it now, it's mine. In retrospect, the disease had been affecting me for at least several years prior to the diagnosis. Simple motor activities like brushing teeth would "get out of synch," even before my

THE THIRD DECADE

attempt to return to the *Andrea Doria*, requiring a concentrated effort to make hand movements that used to be a natural habit. I thought it odd at the time, but nothing serious, as least until my hand started to shake uncontrollably, making it impossible to hide, especially from myself. The trimix training drill of manipulating brass snaps to remove and replace stage bottles had always seemed much harder to do smoothly than it should have been, and my tolerance for cold was negligible. Parkinson's is subtly treacherous; symptoms are initially at the margins of activity, and it progresses with such gradual persistence that it fools a person into thinking for a time that "normal" now is the same as "normal" was six months or six years ago. Because of Parkinson's Disease, I must constantly reassess what constitutes normal physical activity, but with an almost reassuring consistency: being able to do less physically and only with greater effort with each passing month.

In an odd way, the certainty of knowing roughly how things are going to end - assuming I don't get run over by a truck - offers a sense of stability. It forces me to face not just the obvious fact that we are all going to die, but also the more subtle reality that the chances are that I will not get run over by a truck, or come to terms with death in some Hollywood glamorous fantasy of blissful reflection. Dying slowly is hard work, and let's face it: that's what growing old is all about - the daily grind of coping with decreased capabilities, with each new, small mental and physical change and limit seeming to accumulate overnight to gradually block us from activities increasingly associated with being young, or at least younger, all the while knowing how this story must end. Perhaps that's why my *Andrea Doria* experiences have had such a profound influence on me. I was fortunate to have the opportunity to experience a much whittled down preview, but a preview nonetheless, of confronting the changes within that were so thoroughly a part of my being, that I did not even notice their presence. It was my good fortune to learn that I was a fundamentally different person in 2001 than in 1981, and that it was not a fact to lament or celebrate; it simply "be's that way." Returning to the *Andrea Doria* was a wake-up call that helped prepare me for what everyone who avoids that truck accident must learn eventually. Parkinson's picked up pretty seamlessly where my *Andrea Doria* experiences left off, and continues to teach me the same basic lesson, just at a bit of an earlier and faster rate than normal "normal." I'll say it again; there is no substitute for experience.

And what of my old New York dive buddies? Gary Gilligan successfully dove the *Andrea Doria* again the next summer and continued visiting the wreck for years to follow. After his kids got a bit older, Don Schnell - "Hook" - decided to ignore the risk of his lung condition, started diving again, and also returned to the *Andrea Doria*. It was inspirational to see these two old friends persevere and give the finger to the passing years, at least for a time, or maybe they were directing their collective fingers at me, still the dumb college kid who thought a lot, but couldn't get much done on his own; that would be okay too.

Steve Bielenda unfortunately suffered a serious stroke, which I suppose should not have been a great surprise given his fiercely competitive nature and the high stress level he routinely accepted for so many years. Steve stopped diving and sold the *Wahoo* to Hank Garvin who continues to run New York wreck dive charters. Sally Wahrmann still dives and practices her loveable CPA ("Certified Pain in the Ass") skills as a board member of the Women Diver's Hall of Fame. As far as I know, Craig has not made a dive since the late 1980s or early 1990s. I hear from the rest of the old New York crowd only occasionally, mostly Hank Keatts and Dillon, but there is not a doubt in my mind that if I ran into any of my old dive buddies today, we would pick up a conversation as if it were 1984. My New York buddies (actually, Steve and Hook are the only ones still in New York: Gary lives in Connecticut, Sally moved to California, Hank and Dillon retired to Florida, and I lost touch with Craig) are in agreement that the early 1980s were different than anything to follow, special not simply in the dives made, but in the friendships formed.

Five years after my Parkinson's diagnosis, I considered quitting diving altogether, that is, until my son, Jared, turned twelve this past year and I got to watch his instructor hand him a certification card underwater at the end of his final training dive. We were fortunate enough to dive together on a vacation to Maui this past Christmas - yeah, I know; rough life - and

THE THIRD DECADE

I intend to be his dive buddy for maybe one more year in Washington, but only dives to shallow, sheltered spots where we can enjoy the scenery, Jared can build experience, and I can stay within my changing limits. Not anywhere near the cutting edge of diving anymore, it has been a pleasant surprise for me to re-learn that simple diving can be fun.

Witnessing my children pass through each phase of growing up - toddler, Kindergarten, teen - on their way to adulthood cures the soul with pride, frustration, deep love, and profound melancholy as the realization hits that each little persona will soon be gone. Each physical and emotional transition further distances that little boy and little girl, with both soon to disappear forever, never to be seen again except in photo and memory. We are all different people as defined by the march of time and circumstance, and each person we lose in the past is to be mourned and their memory cherished, but that's life. Feeling experience deeply is life.

Parkinson's has taught me to recognize and respect the value of newly framed challenges. That's the real test, and I suppose that it always has been. Stop and smell the anemones, the end of a road is not the point, that part's easy to figure out; enjoy the journey, both with all it has to offer and with what it takes away.

Cheers,
Peter Hunt
Spring of 2011
Whidbey Island

BIBLIOGRAPHY

Books/Periodicals

Carrothers, John. "Accidents are Caused – They Don't 'Just Happen'!" *Titanic Commutator*, Volume 5, Number 1. *Titanic* Historical Society, (1981).

Gentile, Gary. *Andrea Doria: Dive to an Era.* Philadelphia: Gary Gentile Productions, 1989.

Hoffer, William. *Saved.* New York: Summit Books, 1979.

Keatts, Henry and George Farr. *Dive Into History: U-boats.* New York: American Merchant Marine Museum Press, 1986.

Keatts, Henry and George Farr. *Dive Into History: Warships.* Houston: Pisces Books, 1990.

Keatts, Henry and George Farr. *New England's Legacy of Shipwrecks.* New York: American Merchant Marine Museum Press, 1988.

Keatts, Henry and Brian Skerry. *Complete Wreck Diving.* New York: Aqua Quest Publications, 2002.

Moscow, Alvin. *Collision Course.* New York: Putnam, 1959.

Videos

Andrea Doria: The Final Chapter. Peter Gimbel and Elga Andersen. The *Doria* Project, 1984.

SETTING THE HOOK

PETER AND JARED HUNT AFTER A DIVE IN WASHINGTON STATE (2011). PHOTO COURTESY OF GREG MOORE.

Peter Hunt was born in New York and spent six years of his childhood in Athens, Greece where he started diving in 1979. He graduated from Brown University in 1985 before joining the Navy and training as an A-6 Intruder attack pilot. Hunt completed three aircraft carrier deployments to the Persian Gulf, Indian Ocean, and Western Pacific during ten years of military service. After leaving the Navy, Hunt continued to fly as a United Airlines pilot until being diagnosed with Parkinson's Disease in 2005 at age forty-three. Peter Hunt holds a Masters in Strategic Planning for Critical Infrastructure from the University of Washington and lives with his wife and two children on Whidbey Island. His first book, *Angles of Attack* (2002), is a pilot's account of combat operations during the first Gulf War.

SETTING THE HOOK

A deep-sea diver explores shipwrecks and his own character in this gripping scuba memoir.

Hunt (*Angles of Attack: An A-6 Intruder Pilot's War*, 2002) revisits 30 years of shipwreck dives, a pastime whose lugubrious allure is only heightened by his vivid descriptions of the dangers. Chief among these are the hulks themselves, full of ensnaring electrical cables and silt, all of which becomes an impenetrable, disorienting cloud at the kick of a fin; one wrong turn in these pitch-black labyrinths, and a diver can be trapped in a watery tomb. Then there's the sheer physiological challenge of penetrating an alien environment where breathing itself is a high-tech feat rife with fatal glitches. Carbon dioxide can build up to asphyxiating levels; nitrogen first intoxicates and then bubbles out of the blood to cause the bends; even oxygen becomes toxic and induces convulsions. Hunt's well-paced narrative is full of underwater panics, nerve-wracking escapes and rescues that sometimes end in failure and death. He structures it around his dives to the wreck of the Italian cruise ship *Andrea Doria*, which sank in 240 feet of water off Nantucket in 1956—he includes a riveting account of the disaster and the blunders that caused it—and remains a magnet to divers because of its difficulty and wealth of fine china and other loot. Along the way he presents a lucid, engrossing study of the art of diving, introducing readers to the arcane gear, the constant attention to breathing, buoyancy and "situational awareness" the sport demands and the complex decompression routines that make surfacing take twice as long as the dive. Hunt's three decades of *Andrea Doria* excursions also frame an affecting story of maturation and limits, as he ages from a strapping, reckless youth to a more cautious man in physical decline—a transformation that prepares him for the onset of Parkinson's disease with the knowledge that "dying slowly is hard work."

Hunt's taut scenes and meticulous prose will have readers holding their breath, but his saga probes hidden depths as well.

Kirkus Reviews

http://www.kirkusreviews.com/book-reviews/peter-m-hunt/setting-hook/

SETTING THE HOOK

A Diver's Return to the Andrea Doria

This independent but immaculately executed book tells the tale of a diver's relationship with the Italian cruise liner Andrea Doria that sank in 1956 and subsequently became the ultimate scuba adventure, the "Mount Everest of diving." There are many books about the wreck of Andrea Doria, the hunt for "artifacts," the feuds and rivalries between dive boats and personalities, and the 16 divers who have died to-date in the pursuit of this extreme diving adventure. While "Setting the Hook" covers the history of the Andrea Doria and provides much firsthand information on diving expeditions and dives to the wreck, this is also the story of a man's past and present, of assessing and reassessing life.

Author Peter M. Hunt is a retired Navy and commercial pilot who as a young college student in the early 1980s crewed on the dive charter <u>Wahoo</u> that made regular trips to the Andrea Doria. Hunt dove the Doria several times and therefore knows what he's talking about. And the record of his personal friendships with some of the principals of that era adds to the fabric of written Doria history. The special angle of this book is the author's resolve to return to the Andrea Doria almost 20 years later, after life had gotten in the way, taking him away on a military career that included deployment in the first Gulf war. Much has changed in between, from the wreck itself, to deep and technical diving methods and technologies, and to the man himself who is now older and has a wife, career, and children.

SETTING THE HOOK

Setting the Hook (the title refers to crew diving down to the wreck to set the anchor, or "hook") alternates between four eras in time. There's the 1956 sinking of the Andrea Doria when she collided with another ship, the Stockholm, about 50 miles south of Nantucket and 100 miles east of New York City in a freak accident where blame is still being discussed over 50 years later. The slow sinking and rescue efforts (all but 51 of the 1,700 onboard were saved) is well documented, and the Hunt does a fine job describing the tragic event in considerable detail. There are several chapters devoted to the author's dives to the Doria in the early 1980s, recapturing the almost carefree thrill and camaraderie among crew and divers, portraits of the near legendary crew of the dive charter Wahoo (Steve Bielenda, Janet Bieser, etc.), gripping descriptions of dives and penetrations, including some of the tragic accidents and fatalities.

There are chapters describing the intense training the author, by now pushing 40, required in 2000 and 2001 to reacquaint himself with deep diving and getting his trimix certification. Here you find interesting information on trimix training and use, as well as description of at times harrowing dives in the cold Pacific Northwest.

Then there's the section about meeting old friends again, as well as the Wahoo and its crew, now 20 years older, too. The trip back to the Doria — the diver's return — brings a surprise that's very much in line with the overarching theme of this book, that of life's early thrills and drives, the later reflections and efforts at recapturing the magic, dealing with the conflicting priorities of passions, career and family, and finally being able to see everything in perspective.

While there are parallels between Kevin Murray's *Deep Descent* (published in 2001) that's loosely organized along a timeline from the ship's sinking in 1956 to approximately 2000 and primarily an account of the grand wreck and the people drawn to it, in *Setting the Hook* the ship, while a strong presence, is almost incidental. It is about life, youth, growing up, the many directions we're pulled. But unlike many books discussing life's issues, this one never wanders but maintains a laser-sharp focus on Hunt's own personal Mount Everest. Which means you essentially get two books in one; the story of a man's way of dealing with life, and a first rate account of diving the Andrea Doria and all that it involves.

But that's still not all. While hinting that there may be more than meets the eye early in the book, at the end we learn that the author was diagnosed with early-onset Parkinson's Disease in 2005, at the age of only 43. That puts perspective — the main theme of *Setting the Hook* — onto a whole different level.

Like an increasing number of books these days, *Setting the Hook* appears to be self-published. That is generally a bit of a red flag for me. Most of us are still so accustomed to the flawless editing, proofing and layout of professional print publishers that the new era of self-published efforts all too often yields unpleasant surprises.

Hunt, very much to his credit, avoided all the self-publishing pitfalls. *Setting the Hook* is excellently written, well structured, and superbly proofed. There isn't a single error anywhere. That's quite an accomplishment by and in itself. Also nice are almost 40 pictures throughout the book that illustrate the narrative. While they are black and white and thus cannot convey the colors and lack of it underwater, inserting them in the right spots greatly adds to the reading experience.

I greatly enjoyed reading *Setting the Hook*. The book initially caught my attention because of the scuba and Andrea Doria theme (and the great cover). It fully delivered on the diving, adventure and technical fronts, but it was the human angle of the author's very personal journey that elevates this much recommended book above a mere description of events.

— *C. H. Blickenstorfer, scubadiverinfo.com*

CPSIA information can be obtained at www.ICGtesting.com
Printed in the USA
LVOW10s1107061213

364205LV00003B/37/P